PERSPECTIVES ON A DYNAMIC EARTH

PERSPECTIVES ON A DYNAMIC EARTH

T. R. Paton

School of Earth Sciences,
Macquarie University, New South Wales, Australia

with the assistance of J. Clarke

London
ALLEN & UNWIN
Boston   Sydney

Allen & Unwin (Publishers) Ltd,
40 Museum Street, London WC1A 1LU, UK

Allen & Unwin (Publishers) Ltd,
Park Lane, Hemel Hempstead, Herts HP2 4TE, UK

Allen & Unwin Inc.,
8 Winchester Place, Winchester, Mass 01890, USA

Allen & Unwin (Australia) Ltd,
8 Napier Street, North Sydney, NSW 2060, Australia

First published in 1986

British Library Cataloguing in Publication Data

Paton, T. R.
    Perspectives on a dynamic Earth.
1. Earth sciences
I. Title
550   QE26.2
ISBN 0-04-550042-8
ISBN 0-04-550043-6 Pbk

Library of Congress Cataloging in Publication Data

Paton, T. R.
    Perspectives on a dynamic earth.
Bibliography: p.
Includes index.
1. Geodynamics.   2. Paleomagnetism.
3. Plate tectonics.   I. Title.
QE501.P37   1986   551.1   85-26807
ISBN 0-04-550042-8
ISBN 0-04-550043-6 (pbk.)

Set in 10 on 12 point Palatino by Computape (Pickering) Ltd
Printed and bound in Great Britain by
Anchor Brendon Limited, Tiptree, Essex

Only a man who understands science (that is scientific problems) can understand its history; ... only a man who has some real understanding of its history (the history of its problem situations) can understand science.

Karl Popper, *On the theory of the objective mind*

To derive pleasure from the art of discovery the student must be made to relive, to some extent the creative process. In other words, he must be induced ... to experience in his own mind some of the flashes of insight which have lightened its path. This means that the history of science ought to be made an essential part of the curriculum, that science should be represented in its evolutionary context – and not as a Minerva born fully armed. ... The traditional method of confronting the student not with the problem but with the finished solution, means depriving him of all excitement, to shut off the creative impulse, to reduce the adventure of mankind to a dusty heap of theorems. ... Our textbooks and methods of teaching reflect a static, pre-evolutionary concept of the world. For man cannot inherit the past; he has to recreate it.

Arthur Koestler, *The act of creation*

Preface

This book is an attempt to put into practice these precepts of Popper and Koestler as far as they can be applied to the Earth sciences at an elementary level. It is felt that the time is ripe for such a presentation, for the revolution that has taken place over the past 20 years within the Earth sciences has made more people directly aware of the way science works and of the necessity of knowing its history to achieve a full understanding of the problems involved.

Emerging from the revolution has been the immensely unifying and extremely fruitful concept of plate tectonics, and developments leading to its establishment form the core of the book (Chs 4, 5 & 6). However, to see plate tectonics in context, it is necessary to look at what happened before, and this is done in the first three chapters. Chapter 1 is concerned with the development of ideas about the shape, size and mass of the Earth, which led to broad concepts about the Earth's structure and finally to a model of a cooling, contracting Earth, capable of explaining geological history and the major topographic features of the Earth. Chapter 2 goes on to show how even though the acceptance of this idea gradually broke down in the first half of the 20th century, possible alternatives, which are now at the core of plate tectonics, were also rejected. Chapter 3 considers how the debris left by the collapse of the cooling, contracting model were gradually removed and new techniques introduced, which paved the way for the plate tectonic revolution of the 1960s.

The final chapter (Ch. 7) once again considers how the major topographic features of the Earth have been formed, but this time it is done in terms of plate tectonics, rather than of a cooling, contracting Earth. The aim of this is to give some idea of how plate tectonics has developed in the past 20 years, by using as an example one area only, from among the many that have been very profoundly affected by this revolutionary change of ideas. At the same time, it will give some idea of current research problems and also of the explanatory power of plate tectonics.

T. R. Paton
Faculty of Natural Resources
Prince of Songkla University, Hatyai,
Thailand

Acknowledgements

I am grateful to the following individuals and organisations who have given permission for the reproduction of illustrative material (numbers in parentheses refer to text figures):

Figures 1.2 and 1.3 reproduced with permission from I. B. Cohen, *The birth of a new physics*, © 1985 W. W. Norton & Co. and Penguin Books; Thomas Nelson (2.2, 2.3, 2.6, 3.4, 3.7, 3.8, 3.11, 3.12, 3.13, 3.23, 4.2, 5.2); Figures 2.4 and 2.5 reproduced from A. Wegener, *The origin of continents and oceans* by permission of Methuen; Figure 3.5 reprinted with permission from *Geochemica et Cosmochemica Acta*, vol. 10, pp. 230–7, Age of meteorites and the Earth by C. Patterson, © 1956 Pergamon Press; Figure 3.9 reproduced with permission from E. A. Johnson *et al.*, *Terrest. Magn. Atmos. Elect.*, vol. 53, p. 366, © 1946 American Geophysical Union: Harper & Row (3.14); Figures 3.21 and 3.22 reproduced from K. E. Bullen, *An introduction to the theory of seismology*, by permission of Cambridge University Press; Figures 3.24, 4.5, 6.20, and 6.21 reproduced with permission from F. J. Sawkins *et al.*, *The evolving Earth*, © 1963 Macmillan Journals Ltd.; Figure 4.1 reproduced with permission from Heiskamen and Meinesz, *The Earth and its gravity field*, © 1958 McGraw Hill; Figure 4.4 reproduced from H. W. Menard, *Geological Society of America Bulletin*, vol. 66 (1958) with permission from the Geological Society of America; Figure 4.6 reproduced from A. D. Raff and R. G. Mason, *Geological Society of America Bulletin*, vol. 72 (1961) p. 1260, with permission from the Geological Society of America; Figure 4.7 reproduced from V. Vacquier *et al.*, *Geological Society of America Bulletin*, vol. 72 (1961) p. 1251, with permission from the Geological Society of America; Figure 5.3 reproduced from *Proc. R. Soc. Tasmania*, vol. 89, Fig. 10, p. 268, by permission of S. Warren Carey; The Royal Society (5.5); Figure 6.1 reproduced from A. Cox and R. Doell, *Geological Society of America Bulletin*, vol. 71 (1960) p. 736, with permission from the Geological Society of America; Figure 6.2a & b reproduced with permission from A. Cox, *Science*, vol. 144 (1964) Fig. 3, p. 1541, and vol. 163 (1969), Fig. 3, p. 239, © 1964 and 1969 American Association for the Advancement of Science; Figures 6.2c and 6.14 reproduced with permission from F. J. Vine and *Science*, vol. 154 (1966), Fig. 5, p. 1405, © 1966 American Association for the Advancement of Science; Figure 6.3 reproduced with permission from F. J. Vine, *Nature*, vol. 199, p. 948, Fig. 3, © 1963 Macmillan Journals Ltd; Figure 6.5 reproduced with permission from L. R. Sykes, *J. Geophys. Res.*, vol. 72, p. 2137, © 1967 American Geophysical Union; Figure 6.9 reproduced with permission from H. W. Menard, The ocean floor, *Scientific American*, p. 132, © 1969 W. H. Freeman; Figures 6.10 and 6.12 reproduced with permission from J. T. Wilson, *Science*, vol. 150, pp. 483, 485–9, © 1965 American Association for the Advancement of Science; Figure 6.13 reproduced with permission from W. Pitman III, *Science*, vol. 154, p. 1166, © 1966 the American Association for the Advancement of Science; Figures 6.16, 6.18 and 6.23 reproduced with permission from J. R. Heirtzler *et al.*, *J. Geophys. Res.*, vol. 73, pp. 2120, 2123 and 2124, © 1968 the American Geophysical Union; Figure 6.19 reproduced with permission from A. E. Maxwell, *Science*, vol. 168 (1970), Fig. 7, p. 1055, © 1970 the American Association for the Advancement of Science; Walter Sullivan (6.24); Figure 6.25 reproduced with permission from J. F. Dewey, Plate tectonics, *Scientific American*, May 1972, p. 66, © 1972 W. H. Freeman; Figure 7.1 reproduced with permission from Bonnatti and Crone, Oceanic fracture zone, *Scientific American*, May 1984, p. 42, © 1984 W. H. Freeman; Figures 7.2, 7.9, 7.10 and 7.11 reproduced from *Science*, vol. 213, Figs. 4, 5 and 6, pp. 50, 51 and 52 with the permission of Dr Ben-Avraham, © 1981 the American Association for the Advancement of Science; Figure 7.3 reproduced with permission from J. Jackson and R. Muir Wood, The Earth flexes its muscles, *New Scientist*, 11 December 1980, p. 120, © 1980 New Scientist; Figures 7.4 and 7.5 reproduced with permission from P. Molnar and P. Tapponnier, *Science*, vol. 189 (1975), Figs 1 and 2, p. 420, © 1975 American Association for the Advancement of Science; Figure 7.7 reproduced with permission from P. Molnar and P. Tapponnier, *Nature*, vol. 264, p. 322, Fig. 5, © 1976 Macmillan Journals Ltd; Figure 7.8 reproduced with permission from Dr Coney, *Nature*, vol. 288, p. 330, © 1980 Macmillan Journals Ltd.

Contents

List of tables

CHAPTER ONE

A cooling, contracting Earth

The ability to explain the macrotopography of the Earth, in terms of it being a cooling, contracting planet, obviously requires a great deal of scientific sophistication, and the history of how this was achieved by the latter half of the 19th century is the subject of this first chapter.

Initially, the Earth was considered to be flat, for this equated with everyday experience of people who lived in small communities and whose known world measured a few kilometres in each direction. However, what appeared to be such a commonsense view had inherent logical difficulties. Where did the Earth end? Could it go on for ever? The concept of an infinite Earth was an uncomfortable one. But then so was the alternative view that the Earth was a flat body of finite size, for then the problem was one of explaining how the Earth was supported and prevented from falling through space.

The problem was resolved only when a move was made away from the concept of a flat earth, and this was caused by the necessity of explaining certain facts that became apparent as man travelled more widely across the surface of the Earth. Thus it was observed that on travelling northwards some stars disappeared below the southern horizon and others appeared over the northern horizon. This was inexplicable in terms of a flat Earth, but could be explained if the Earth was a cylinder with an east–west axis; that is, the Earth's surface was curved in a north–south direction. One of the implications of this idea was that ships heading out to sea in a north or south direction would disappear from view hull first as they moved over the curvature of the cylinder, while ships travelling east or west parallel to the axis would remain totally visible, but smaller and smaller, until they disappeared from view. However, observation showed that ships disappeared from view hull first no matter in what direction they travelled, which was explicable only if the Earth was curved equally in all directions; in other words, it was spherical rather than cylindrical. Astronomical observations of lunar eclipses supported this idea. It was reasoned that such eclipses were caused by the shadow of the Earth being cast by the Sun on the Moon. As this shadow is always circular in outline, no matter what the position of the Sun, Earth and Moon, the Earth must be a sphere, for only a sphere is capable of producing this effect.

This suggestion of a spherical Earth, made in about 450 BC, resolved the problem of the 'end' of the Earth without resorting to infinity. A sphere is of finite size but has no end; it is finite without lateral boundaries. The spherical Earth also resolved the problem of 'down', for in the terms of this idea 'down' means in the direction of the centre of the sphere, and so to say that things 'fall downwards' means that they fall towards the centre of the Earth. In addition, the spherical Earth removed the need to support the Earth because every part had already fallen as far as possible towards its centre and indeed this was the reason for its spherical shape.

By about 350 BC, the spherical shape of the Earth had become firmly established, as this model conformed with all observations and resolved all the problems of the flat Earth. It should, however, be noted that there was no direct proof of the Earth's sphericity until the circumnavigation by Magellan's fleet in 1522, almost 1800 years later.

With the shape of the Earth established, the question of its size became meaningful. It could be observed that the Earth was not very large, for the Earth's shadow on the Moon during eclipses is always markedly curved and the star pattern changes considerably in moving only a few hundred miles north or south across the Earth's surface. The first person to suggest a quantitative answer to this problem was Eratosthenes (276–196 BC), who was aware that on midsummer's day (21 June) the noon Sun was reflected by the water in a deep well near Syene (modern Aswan), which meant that the Sun was directly overhead. On the same day, at the same time, 820 km to the north at Alexandria, it was known from the shadow cast by a pillar that the Sun was 7° 20' south of the vertical (Fig. 1.1). Assuming the Sun's rays are parallel, it follows that the radii connecting the ends of this 820 km long arc with the centre of the Earth subtend an angle of 7° 20'. From this it can be calculated that the 360° circumference of the Earth amounts to some 40 250 km and its diameter to 12 900 km. These figures produced some 2100 years ago compare amazingly well with the most recent measurements of 38 402 km for the Earth's circumference and 12 768 km for its diameter.

Thus by 200 BC the Earth's approximate shape and size had been established, but there was a gap of over 1800 years before its mass was successfully determined, for the problem was insoluble until it was recognised that the Earth was just another planet in the Solar System and that there was no fundamental difference between the physical processes that occur on the Earth and those taking place on the other planets.

This recognition was prevented by the development of two interrelated concepts. The first was the physics of Aristotle that divided the Universe into the Heavens and the Earth, each composed of different materials behaving in different ways. The Heavens were perfect and unchanging, and so all movement was perfect and hence circular; while materials on Earth were imperfect and corruptible. This raised a conceptual barrier

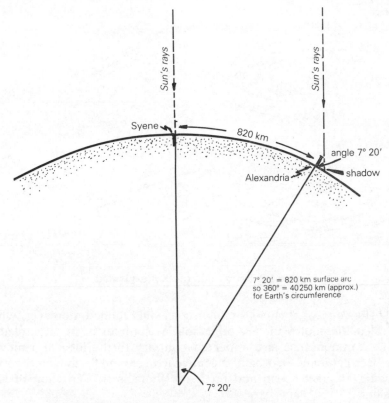

Figure 1.1 Eratosthenes and the size of the Earth.

between terrestrial and planetary physics. In addition, the circular motion of the heavenly bodies around the central, stable Earth (Aristotle's geocentric model) was further developed by Ptolemy of Alexandria in the 2nd century. He interpreted the observed irregular motion of planets as being due to a combination of circular motions that resulted in a very elaborate system of calculations by which he was able to predict planetary positions successfully. These two concepts were not seriously challenged until the 16th century for, although based on false assumptions, they satisfied the needs of most men, provided a framework that explained most simple observations and had a utilitarian value in predicting planetary positions.

In 1534 Copernicus revived the ideas of the Greek Aristarchus and replaced the geocentric system with the heliocentric (or Sun-centred) system, in which the Earth orbits the Sun, as do the other planets, and also rotates on its axis, while being orbited by the Moon. Using this model, he was able to give a much simpler explanation of the daily movement of the Sun, Moon, planets and stars and of the apparent retrograde motion of planets such as Mars and Venus (Fig. 1.2). In addition, he was able to calculate the relative distance of each planet from the Sun and its orbital

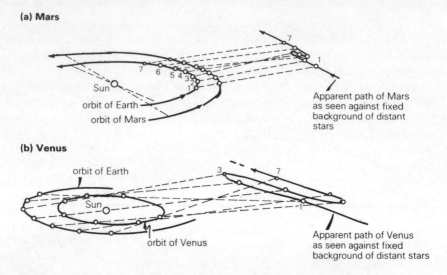

Figure 1.2 Retrograde motion explained.

period. The concept of circular motion was still retained, however, which made calculations of planetary orbits just as complex as in the Ptolemaic system. Commonsense also appeared contrary to the idea of a moving Earth, for to complete an orbit of the Sun each year the Earth would have to be moving at great speed and on Earth there is no sensation of such movement.

However, Copernicus had lit a fuse, albeit a slow-burning one, for the explosion did not occur for over 60 years. This took place in 1609 when Galileo published the *Starry message*, which gave results of the first application of the telescope to astronomical problems. At one stroke, this work established the reality of Copernicus's heliocentric Solar System, which could no longer be dismissed as being merely mathematical speculation, and at the same time almost completely refuted the basis of the Aristotelian–Ptolemaic system.

Galileo demonstrated that the Moon had mountains and valleys and even measured the height of the mountains. Such evidence obviously showed that the nearest heavenly body was very similar to the Earth. Observations also showed that some of the heavenly bodies were not as perfect or as unchanging as had been imagined; Saturn appeared to have a pair of 'ears' that sometimes changed shape and even disappeared (these were the rings of Saturn and could not be completely resolved with the telescopes then available), while the Sun's surface was observed to be spotted, hardly to be expected of an unblemished celestial object.

That the Earth behaved as a planet was also demonstrated by the observation that the dark side of the Moon was still illuminated to a certain extent. This illumination could not be attributed to a source internal to the

Moon, or to starlight, for it did not continue during eclipses. It could only be attributed to the light radiated from the Sun and reflected by the Earth onto the Moon (called 'earthshine'). In other words, the Earth reflected sunlight and shared this characteristic with the other planets.

There was also direct confirmation of the Copernican heliocentric model. Observations of Venus showed that its apparent size varied from being largest when a crescent to smallest when full, which was explicable only if Venus orbited the Sun, presenting its fully illuminated face to the Earth only when furthest away (Fig. 1.3).

In the case of Jupiter, Galileo discovered that it was orbited by four moons. This demonstrated the actual occurrence of a Copernican-style system and made more plausible that part of the Copernican model which had provoked the greatest resistance: that the Moon orbited the Earth while the Earth orbited the Sun. Here was an example of a planet being orbited by four moons, while it in turn orbited the Sun.

The work of Galileo undoubtedly established that the Earth was a planet, rotating on its axis and orbiting the Sun together with the other planets. There remained, however, the problem of the paths of the planets round the Sun. These were still seen in terms of circular orbits, although this did not fit the observational facts very well. This problem was resolved in 1609 by Kepler, following years of work during which he found every kind of circular motion to be unsatisfactory. He finally showed that the orbits were not circles, but ellipses (Fig. 1.4a), with the Sun located at one focus of the ellipse. He then demonstrated that a line connecting an orbiting planet to the Sun will sweep out equal areas in equal intervals of time (Fig. 1.4b), so that the closer a planet is to the Sun the faster it moves along its orbital path. Furthermore, he established that the square of the relative orbital period of any planet is equal to the cube of its relative distance from the Sun (Table 1.1). This was either a most amazing coincidence, or else there was some fundamental reason for these relationships.

Aristotelian physics, the basis of which was destroyed by Galileo, had been developed to account for the behaviour of objects in relation to a static

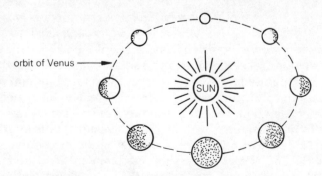

Figure 1.3 The phases of Venus as related to its apparent size as viewed from the Earth.

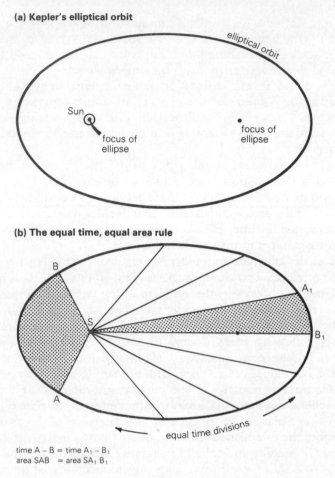

(a) Kepler's elliptical orbit

(b) The equal time, equal area rule

time A – B = time A₁ – B₁
area SAB = area SA₁ B₁

Figure 1.4 Kepler's contribution to the theory of planetary motion.

non-mobile Earth. The acceptance of the Earth as a planet rotating on its axis and orbiting the Sun meant that a new mobile Earth physics was required. Galileo attempted this in 1638 when, in his *Discourses and demonstrations concerning two new sciences*, he established principles that could be expressed mathematically for the motion of falling bodies on a mobile Earth.

It had been recognised for some time that the natural motion of a falling body is continuously accelerated. Galileo's great achievement was to quantify this acceleration. He established that the distances traversed by a body falling from a position of rest, in successive equal time intervals, were in the ratio 1 : 3 : 5 : 7 and so on. Thus the ratios of the total distances traversed by the end of each successive equal time interval are 1 : 4 : 9 : 16 .... The total distance traversed therefore is proportional to the square of the elapsed time, a rule applicable to any body falling towards the surface of

Table 1.1
The orbital periods of the planets and their distances from the Sun relative to those of the Earth.

	Mercury	Venus	Earth	Mars	Jupiter	Saturn
orbital period	0.24	0.614	1.0	1.88	11.68	29.457
(orbital period)^2	0.058	0.38	1.0	3.54	140	868
distance from Sun	0.387	0.723	1.0	1.524	5.203	9.539
(distance from Sun)^3	0.058	0.38	1.0	3.54	140	868

the Earth. However, this gave only a relative measure and it was not until Huygens invented an accurate clock almost 20 years later (1657) that the acceleration due to free fall at the Earth's surface could be quantified absolutely as being 9.8 m s^{-2}.

Despite this great advance, Galileo could not extend his system of terrestrial dynamics (in which natural motion was in a straight line) to the rest of the Universe. He still thought that natural motion outside of the Earth was circular, as he refused to accept the reality of Kepler's elliptical orbits. It required the genius of Newton to achieve this breakthrough in his *Principia* (1687). Newton fused together different lines of evidence to give a comprehensive system for the movement of bodies throughout the Universe, in all places and at all scales. He did this by boldly stating that the force responsible for the way in which objects fall towards the surface of the Earth (Galileo's dynamics) and the force that holds planets and satellites in their orbits (Kepler's work) is the same. In one sweep his universal force of mutual attraction, or gravitation, completely removed the debris of the Aristotelian–Ptolemaic systems, by postulating a mechanism that had universal applicability.

Newton's achievement stemmed from considering the force that holds the planets in their orbits. He realised that the elliptical shape of the planetary orbits and the fact that planets move faster the closer they approach the Sun imply that they are constantly experiencing an acceleration towards the Sun. He showed that this acceleration must be proportional to the inverse square of the planet's distance from the Sun. Furthermore, from Kepler's observation of the proportional relationship between the square of the orbital period and the cube of the mean distance from the Sun, it can be shown that the constant of acceleration is the same for all planets; that is, they would all have the same acceleration if they were at the same distance from the Sun. Since acceleration is the result of a force, the Sun must exert a force on the planets. Newton suspected that this was the same force of attraction (or gravity) that the Earth exerted upon objects at its surface.

To establish this connection he examined the Earth–Moon system. It was known that the Moon moves at 1020 m s^{-1} and Newton calculated that to maintain its orbital path around the Earth the Moon must fall towards the Earth a distance of 0.0027 m each second. Assuming the force of gravity did extend to the Moon, the acceleration of the Moon towards the Earth could be calculated. As an object at the Earth's surface has an acceleration due to gravity of 9.8 m s^{-2} (from Huygen's work), and the Moon is 60 Earth radii distant, then by the inverse square law the Moon would have an acceleration towards the Earth of $(9.8/60^2)$ m s^{-2}. This works out to be 0.0027 m s^{-2}, showing that gravity does indeed extend as far as the moon.

Having demonstrated that the Earth's gravity holds the Moon in its orbit, it could then be inferred that the Sun's gravity holds the planets in their orbits. This being so, it was reasonable to conclude that the rule for earth-bound bodies, that every action is accompanied by an equal and opposite reaction, was also applicable to the Sun and the planets. Thus every planet is attracted by the Sun but attracts the Sun with equal force. Since force is proportional to mass multiplied by acceleration and a planet's acceleration is inversely proportional to the square of its distance from the Sun (that is, its mean orbital radius), the force exerted by the Sun on any planet must be proportional to the mass of the planet divided by the square of the mean orbital radius, or

$$\text{force exerted by Sun on planet} \propto \frac{\text{Mass of planet}}{(\text{Mean orbital radius})^2}$$

At the same time the planet is attracting the Sun with a force that must be proportional to the mass of the Sun. Therefore, the force of gravitational attraction between the Sun and the planet (F_{sp}) is proportional to the mass of the Sun (M_{s}) multiplied by the mass of the planet (M_{p}) divided by the square of the mean orbital radius (r), or

$$F_{\text{sp}} \propto \frac{M_{\text{s}} M_{\text{p}}}{r^2}$$

It was immediately realised that this force of gravitational attraction was not just acting between bodies of planetary size but was responsible for a force of attraction between *any* two bodies. Thus the formula could be rewritten as

$$F \propto \frac{M_1 M_2}{d^2}$$

where M_1 and M_2 represent the masses of any two bodies and d is the distance between them. This could then be changed into

$$F = \frac{G M_1 M_2}{d^2}$$

8

where G, the constant of proportionality, is invariable and is known as the universal constant of gravitation.

The use of these expressions enabled the obstacle that had stopped progress in 200 BC to be overcome; now it was possible to determine the Earth's mass, for it followed that

$$\text{gravitational attraction between a small object and the Earth} \propto \frac{\text{mass of the Earth}}{(\text{distance from Earth's centre to small object})^2}$$

or

$$A \propto C$$

and

$$\text{gravitational attraction between a small object and a large object at the surface of the Earth} \propto \frac{\text{mass of large object}}{(\text{distance between objects})^2}$$

or

$$B \propto D$$

Thus

$$A : B = C : D$$

A is known, for this is the acceleration of a freely falling body, which is 9.8 m s^{-2} at the Earth's surface. In C the denominator is the Earth's radius and is known. In D both of these facts can be determined. Therefore, if a method could be developed to determine B, the attraction between a small and a large object, the way would be open for evaluating the only remaining unknown, the mass of the Earth.

Throughout the 18th and 19th centuries various methods were suggested to solve this problem. One used plumb-bobs as the small object and mountains as the large object. This was first applied by Bouguer in 1737 when in Peru engaged in the measurement of the length of a degree of latitude. He placed one plumb-bob on the slopes of an isolated mountain (A in Fig. 1.5) and one some distance from the mountain (B). The angle of elevation of a particular star when at its zenith was observed from both sites. The angle of elevation was, of course, measured with reference to the vertical given by the plumb-bob. At the mountain station, however, the plumb-bob would be attracted by the mass of the mountain and so the star zenith angle would differ from that expected. This expected value can be calculated from the angle subtended at the Earth's centre by radii passing

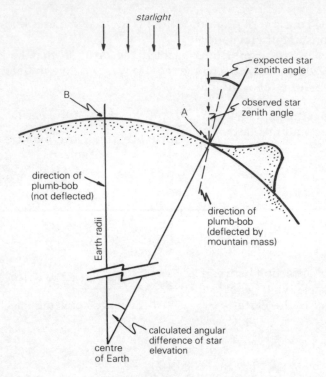

Figure 1.5 Bouguer's experiment to determine the mass of the Earth by plumb-bob deviation.

through these two stations, for these radii would give the true vertical, i.e. the directions taken by plumb-bobs at the Earth's surface if no mountains were present. The difference between the measured and calculated angles is then a measure of the deviating effect of the mountain on the plumb-bob. The force necessary to cause this deviation could then be calculated. Knowing the volume and density of the rocks making up the mountain, everything is known in the expression given previously except for the mass of the Earth and this can then be determined (see box).

Bouguer's reasoning was impeccable, but the execution of the plan was faced with immense physical and instrumental difficulties. The mountain selected was the volcanic peak of Chimborazo, rising 6000 m above sea level, so that station A was above the snowline and B was in a cold desert region. Given these conditions it is remarkable that the result was only 5–6 times the presently accepted value. Bouguer realised that this was very much a first attempt that needed considerable refinement.

Thirty-seven years later (1774) the experiment was repeated by Maskelyne, the Astronomer Royal. The mountain selected this time was the 1080 m high Schiehallion near Loch Rannoch in Perthshire, Scotland. This was a much more manageable object than Chimborazo having a short

Determining the mass of the Earth

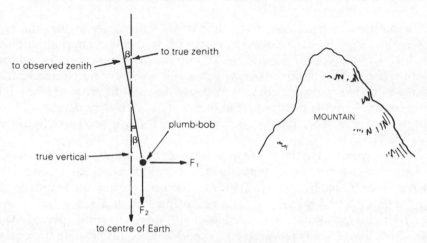

The force pulling the plumb-bob towards the mountain is

$$F_1 = \frac{GM_m m}{d^2}$$

where
M_m is the mass of the mountain, determined from knowledge of its shape and density, m is the mass of the plumb-bob, d is the distance from the plumb-bob to the centre of gravity of the mountain, and G is the gravitational constant.

The force pulling the plumb-bob towards the centre of the Earth is

$$F_2 = \frac{GM_e m}{R_e^2}$$

where M_e is the mass of the Earth, m is the mass of the plumb-bob, and R_e is the radius of the Earth.

The angle of deflection β of the plumb-bob, equal to the angle between the true and observed zenith, is given by resolution of the two forces; thus

$$\tan \beta = \frac{F_1}{F_2} = \frac{GM_m m}{d^2} \frac{R_e^2}{GM_e m}$$

Here G and m cancel out, thus

$$\tan \beta = \frac{M_m}{M_e} \frac{R_e^2}{d^2}$$

or

$$M_e = \frac{M_m}{\tan \beta} \frac{R_e^2}{d^2}$$

from which the mass of the Earth can be obtained.

[Prepared by F. Williams]

east–west ridge and steep north and south slopes. The topography and the nature of the bedrock of the mountain were also very much better known than for Chimborazo.

The method used differed from that used by Bouguer in that the two observation stations were located one on each side of the mountain (on the north and south slopes), so that the plumb-bob deviations reinforced one another (Fig. 1.6). On this occasion the result was very much nearer to a value acceptable at the present day. Because the mass figure is so large, it is more comprehensible to express it in terms of its equivalent density, which in this case was 4.5 g cm^{-3} compared with today's accepted figure of 5.5 g cm^{-3}.

Another approach to the problem of measuring the mass of the Earth was by using a pendulum as a small object, for the rate at which a pendulum swings is controlled by gravity. The first workers adjusted the length of the pendulum to get a 1 s swing at each recording station. Later an invariable (fixed length) pendulum was used and the time of swing (period) was measured. These time measurements could be used to calculate the equivalent length of the pendulum required to produce a 1 s swing at that place. Once again it was Bouguer who pioneered this method in Peru by determining the length of a 1 s pendulum at three elevations. These were (1) at Quito, with an elevation of 2862 m, (2) on the summit of Pichincha, which rises behind Quito to a height of 4750 m above sea level, and (3) on Inca Island on the River Esmeraldo, which is not more than 60 m above sea level and is on the same latitude as Quito.

A decrease in gravity with altitude was observed (Table 1.2), but this decrease was less than would have been expected if the measurements had been taken at the same elevation with no mountains present, i.e. the value that would be calculated by application of the inverse square law. The difference between these figures gives the attractive force exerted by the mass of Pichincha, which again enables the mass of the Earth to be calculated. Just as in his plumb-bob deviation work, practical and instrumental difficulties once again made Bouguer's answer far too high. However, as Bouguer remarked, both sets of experiments in Peru had

Table 1.2
Results of Bouguer's experiment in the use of pendulums to determine the mass of the Earth.

Station	Elevation (m)	Length of a 1 s pendulum (1/12 inch)	Fraction less than at sea level	Fraction expected by inverse square law
Pichincha	4750	438.69	1/845	1/674
Quito	2862	438.88	1/1331	1/111
Inca Island	60	439.21	—	—

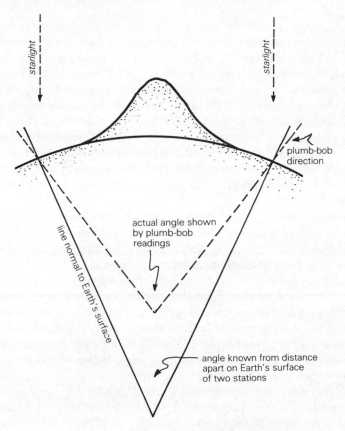

Figure 1.6 Maskelyne's experiment.

demonstrated that the average density of the Earth was greater than the density of the Andes and consequently the Earth was neither hollow nor full of water as had been maintained by a considerable number of people.

By the end of the 18th century, experiments using both plumb-bobs and pendulums had shown that the density of the Earth was between 4 and 7 g cm^{-3}. However, because natural masses were being used, there were necessarily many unknowns, such as the lack of detailed knowledge of the composition and therefore of the density of the Earth's crust. Consequently the results could only be regarded as rough approximations, so attention was turned to the development of a technique for measuring the attraction between accurately known masses, a task that could be handled in a laboratory situation.

This task was accomplished by Cavendish in 1798, using apparatus devised by the Rev. John Michell, who died before he could make use of it and which then came into Cavendish's hands. The basis of the experiment was quite simple, as shown in Figure 1.7. Two small lead balls, 5 cm in diameter, were hung by short wires from the end of a long rod that was supported from its middle by a long thread. Two larger lead balls, 30.5 cm in diameter, were hung down from a cross bar so that they were close to the smaller lead balls and were on either side of the rod that supported them. The attraction between the larger and smaller balls would tend to turn the rod. After determining the movement in one direction, the large balls were placed on the opposite side of the small balls and the movement in the other direction determined. The great difficulty about this experiment was the exclusion of outside influences, for the attractive force was so small that chance air movements could ruin the whole experiment. It says much for Cavendish's ability that such good results were obtained. From 29 measurements he obtained a mean value for the Earth's density of 5.448 g cm^{-3}.

Thus, after a hiatus of 2000 years, the problem of the Earth's mass was resolved. Also during the 17th and 18th centuries the spherical shape of the Earth was reassessed. Newton realised that the rotating Earth of Copernicus could not have the spherical shape proposed by the Greeks, for the Earth's speed of rotation would cause the equator to bulge a little and the poles to flatten. Newton calculated that the equatorial diameter would be 34 miles greater than the polar diameter, a relatively small amount but sufficient to be of considerable importance in gravity measurements. This explained why a 1 s pendulum had to be slightly longer in Paris than on the equator in French Guinea, for gravity is stronger in the former because Paris is marginally nearer to the centre of the Earth.

One consequence of the distortion of the Earth from a spherical shape is that the length of a 1° meridian arc would be less nearer the equator than nearer the pole. Some 50 years after Newton, in 1735, Cassini measured 1° meridian arcs in both northern and southern France. He claimed that the northern one was shorter than the southern one (the opposite to Newton's

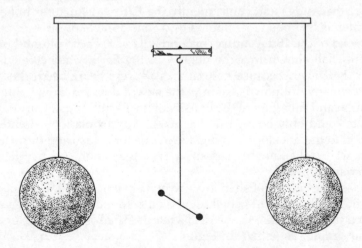

Figure 1.7 Cavendish's experiment.

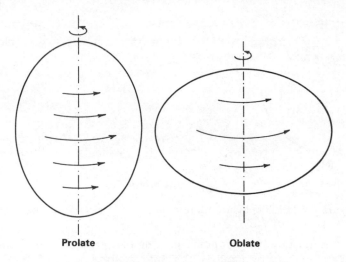

Prolate　　　　　　　　　　　**Oblate**

Figure 1.8 Spheroidal shapes.

prediction) and hence the Earth was a prolate rather than an oblate spheroid (Fig. 1.8).

There was doubt, however, about the accuracy of these results, for the two measurements were made relatively close together and their precision was probably not sufficient to detect any small real difference. In 1736, to resolve this matter the French Academy of Sciences sent expeditions to Lapland under Maupertuis and to Peru under Bouguer (referred to previously), both to measure the length of a degree. The results from these widely separated locations confirmed Newton's prediction that the Earth is an oblate spheroid.

By combining these facts about the Earth's density and its rotational shape, it was found to be possible to make certain deductions about deep Earth structures. Accepting an average of 5.5 g cm^{-3} for the Earth's density and 2.5 g cm^{-3} as the average density for surface rocks, it follows that much denser material must occur somewhere within the Earth. The distribution of this denser material could be deduced by considering certain rotational characteristics of the Earth. The equatorial bulge of the oblate spheroidal Earth causes the Earth to react differentially to the gravitational attraction of the Sun and Moon and is reflected in a slight but measurable wobble, or nutation, in the Earth's rotational axis. The mass responsible for this nutation is extremely small compared to the mass responsible for the average density of the Earth, yet no nutational effect can be ascribed to it. This must mean that the mass is centrally and symmetrically located so as not to be differentially affected by the Sun and Moon. Thus of the three models of mass distribution shown in Figure 1.9, only (c) agrees with the rotational characteristics of the Earth.

It was one thing to establish that the Earth had a dense core but quite

15

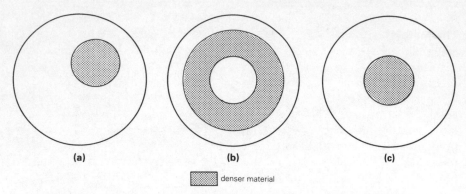

(a) (b) (c)

denser material

Figure 1.9 Possible distribution of denser material within the Earth.

another to determine its composition. The complete inaccessibility of the core made argument by analogy the only method available to tackle this problem. In view of this difficulty, it is remarkable to find that, as early as 1794, E. F. F. Chladni suggested that iron meteorites (see box) could be of similar composition to the core of the Earth, an analogy that is still considered to be valid.

Meteorites

Meteorites are solid bodies derived from outside the Earth, but from within the Solar System. The most likely source is the asteroid belt, which lies between Mars and Jupiter. The asteroids may be either the fragments of a former planet or the pieces of raw material of a planet that never finished assembling. In terms of composition, meteorites are either 'irons', i.e. nickel–iron alloys, or 'stones', i.e. mainly silicate minerals similar to those of the ordinary igneous rocks of the Earth.

Thus, by 1800, the broad outlines of the Earth's shape, size and mass had been blocked out and subsequent developments were concerned with trying to work out some of the details. This was done first of all for the near-surface zone of the Earth, by the utilisation of the plumb-bob methods previously used in determining the Earth's mass, and arose from the measurement of a great meridian arc by the Indian Survey between latitudes 8° 09' and 29° 30' N during the 1840s. The more northerly part of this survey approached the Himalayas and it followed, from Maskelyne's work of the previous century, that this immense mountain mass would cause the plumb-bobs of the surveying instruments to deviate from the true vertical. It was essential that the surveyors were aware of the magnitude of this effect so that the necessary corrections could be applied to their results. The method devised to solve this problem consisted of taking two stations 600 km apart towards the northern end of this meridian arc and measuring the difference in latitude between them by the standard astronomical method and then by the much more cumbersome trigonometrical method. The astronomical method was of course similar to that used by Bouguer and Maskelyne and hence plumb-bob deviations would be involved in its use. However, this was not the case for the trigonometrical method which involves the determination of triangles beginning from an accurately measured base, so that the measured angles are almost totally confined to the horizontal plane, apart from slight departures due to the topography, and hence the instrumental readings are not affected by the attractive force of mountains. The difference in latitude between the two stations by the trigonometrical method was found to be 5° 23' 42.294" and by the astronomical method 5° 23' 37.058", a difference of 5.236 seconds of arc. This result enabled the surveyors to apply the necessary correction to their work, but for others of the survey team it afforded an opportunity to do much more.

By 1854, eight years after those results were available, J. H. Pratt, the Archdeacon of Calcutta, had developed a method that enabled him to calculate the attractive force that the Himalayas should have exerted and found it to be three times greater than that needed to give the surveyors their results. It was as though the mountains were not pulling their full weight. As no fault could be found either in the surveyor's work or that of Pratt, it followed that the model of the structure of mountains on which the calculations were based had to be wrong.

This model regarded mountains as being merely an extra mass of rock material superposed on a rigid substratum (Fig. 1.10a). One possible explanation for the discrepancy was that the mountains were hollow and indeed in the previous century Boskovitch, in 1755, had suggested that mountains were swellings or blisters raised by the Earth's internal heat. Work such as Bouguer's on the Earth's density had since shown, however, that this could not possibly be correct.

Another possibility was that the lower-density rocks of the mountains projected down into the underlying higher-density material, displacing some of it to give a lower-than-expected gravitational attraction. In 1855 G. B. Airy, the Astronomer Royal, suggested that this displacement was the result of the substratum acting as a fluid to some extent. The overlying solid surface, or crust, being less dense could thus be considered as floating on this substratum. If the crust is thicker in some places, as it may be in mountainous areas, the base of the crust would sink into the underlying material until the buoyancy provided by this material equalled the mass of the mountain (Fig. 1.10b). Thus the surface protuberance of mountains may be compensated for by a 'root' protruding into the denser material beneath. Conversely, topographic depressions such as those of the ocean basins would be underlain by an upward bulge of the denser subsurface material

17

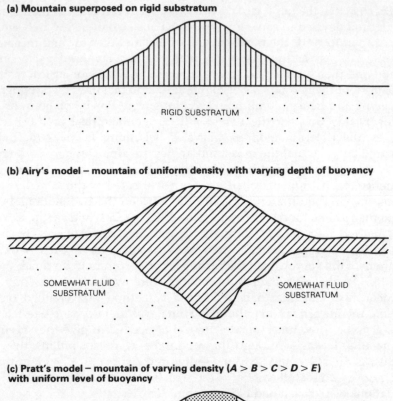

(a) Mountain superposed on rigid substratum

RIGID SUBSTRATUM

(b) Airy's model – mountain of uniform density with varying depth of buoyancy

SOMEWHAT FLUID
SUBSTRATUM

SOMEWHAT FLUID
SUBSTRATUM

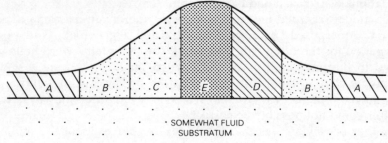

**(c) Pratt's model – mountain of varying density ($A > B > C > D > E$)
with uniform level of buoyancy**

A   B   C   E   D   B   A

SOMEWHAT FLUID
SUBSTRATUM

Figure 1.10 Mountains and their foundations.

beneath the much thinner crust. So Airy's model stated that the Earth's crust is of uniform density, but of variable thickness and that the lower boundary of the crustal layer is inverted with respect to the surface topography.

Airy's model, however, was based on the Himalayan results only. Archdeacon Pratt's work over the next 20 years extended gravity observations to a much wider area. He demonstrated that, if the crust was considered as being uniform in density, then gravity was more than expected in coastal areas compared to continental interiors and much more than expected on oceanic islands. He interpreted these observations as meaning that the floors of ocean basins are composed of material denser than continental rocks, for example basalt. His crustal model therefore (Fig. 1.10c) consisted of a layer of variable density, such that less dense materials give rise to mountains, by floating higher, and more dense materials to ocean basins, by sinking lower, with the base of the crust at a uniform depth.

The original statement by Pratt was highly confused. The model discussed here is a much clarified form of the original that was devised by the American Heyford in 1910 and generally referred to as the Pratt–Heyford model. The term 'isostasy' to cover ideas of crustal flotation was coined in the 1890s by the American geologist Dutton.

These isostatic models of the Earth's crust and in particular Pratt's idea of a basaltic ocean crust were incorporated into a generalised Earth model proposed by J. D. Dana. This model not only gave an integrated view of geological processes but also explained the development and distribution of the major features of the Earth's surface. It was based on the idea of an initially molten Earth, which was in the process of cooling, contracting and solidifying. Dana postulated that at the time of the initial solidification of the surface, large areas had granitic composition while other areas were basaltic. Since granite was known to melt at a higher temperature than basalt, the granite would have solidified before the basalt, but the surface of both would have remained at the same level (Fig. 1.11a).

After the basaltic material solidified, its surface would have sunk below the level of the granite owing to contraction on cooling and so the differentiation between oceanic depressions and continental plateaux would have begun (Fig. 1.11b). With continued cooling, water would accumulate in these depressions, further cooling them and hence aiding the deepening process. This process was thought to account for the greater part of the present difference between the mean depth of the ocean basins and the mean elevation of the continents. The end result of this process was a series of granitic plateaux separated by basaltic depressions and all covered by the waters of an initial world-wide ocean (Fig. 1.11c).

Subsequent Earth history involved the gradual development of the dry land areas of the continents and their mountain ranges from the initially submerged granitic plateaux. This was explained in terms of the further contraction of the Earth as it continued to cool beneath the already solid crust. The crust would have to accommodate itself to this contraction, leading to the development of lateral compressive forces within the crust.

Because of the difference in level between the depressions and the plateaux, the basaltic crust of the former would act as a lever against the granitic crust of the latter, so that the lateral pressure would be directed from the oceanic depressions towards the continental plateaux (Fig. 1.12a). Initially such pressures would have caused a flexing of the plateaux,

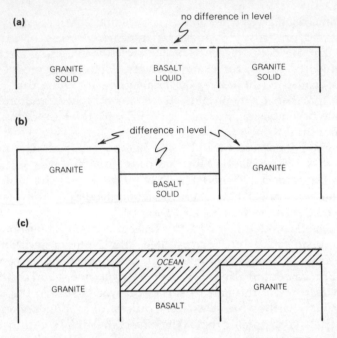

Figure 1.11 Dana's model of initial development of the Earth's surface.

producing broad swells rising above sea level and equally broad submerged depressions. This initiated erosion and sedimentary material was transported from the raised areas and deposited in the depressions (Fig. 1.12b).

The overall process was continued by isostatic adjustment (the vertical movement of a section of the Earth's crust in response to an increase or decrease in load) consequent on the erosion and deposition and combined with continuing lateral pressure. As the sediments in the basins were covered and depressed to greater depths, they entered high-temperature zones that greatly weakened or even melted them (Fig. 1.12c). This meant that the shrinkage-generated lateral pressure could be relieved by intense folding and fracturing within the zone of weakening, leading to the formation of fold mountain belts (Fig. 1.12d). After folding, the continued application of lateral pressure from the ocean basins initiated the whole process again.

This model could explain the formation of continents and ocean basins and within the continents the difference between fold mountains, plainlands and the continental shelf. It was particularly applicable to North America, which Dana regarded as illustrating most simply and perfectly his laws of Earth genesis, for it is bounded on all sides by mountains that face oceanic basins. The conditions under which lateral pressure developed from the ocean basins were therefore simple and continental evolution had been regular and systematic. Thus the largest and loftiest mountain chains,

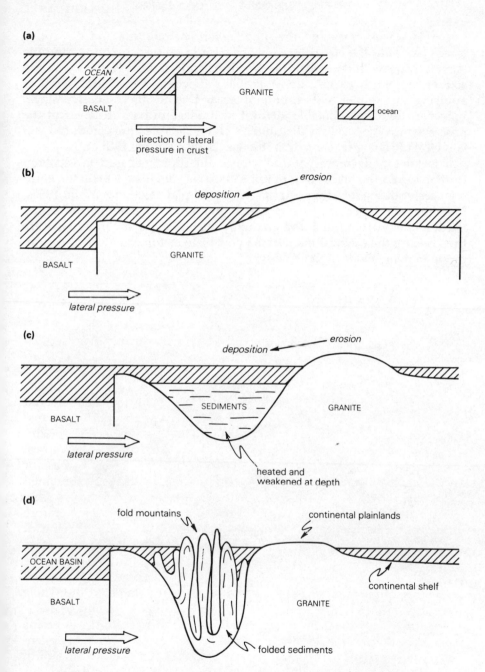

Figure 1.12 Subsequent continent and fold mountain development.

the greatest volcanoes and the other results of uplifting and disruptive forces face onto the largest oceans. This could be seen by comparing the Appalachians with the size of the North Atlantic and the Rockies with the size of the North Pacific. Considerable difficulties were encountered in applying this model to Europe and Asia. Despite these shortcomings, however, this was a notable attempt at unification both in terms of the processes involved and the distribution of the macrofeatures produced and is typical of geological thought in the last 20 years of the 19th century.

In fact this model represents the acme of development based on a cooling, contracting Earth. An equivalent unification was not to be seen in the Earth sciences until the development of the idea of plate tectonics in the 1960s. The intervening period is one in which the foundations of the contracting Earth model were refuted and gradually abandoned and conditions were prepared for the establishment of the new plate tectonic model by quantification in many fields of Earth science.

CHAPTER TWO

The cooling, contracting Earth refuted and a possible alternative rejected

The idea of a cooling, contracting, initially molten Earth has a history in geological thought extending back to the early 17th century. It continued to be important in the second half of the 19th century, because it had been given quantitative teeth by Lord Kelvin, arguably the most prestigious physicist of that time.

Kelvin was involved in developments in thermodynamics in the 1840s and 1850s. These showed that in a closed system the total amount of energy is constant and that every energy transformation (that is, every physical process) decreases the energy available for useful work, so that the system gradually runs down. Applying this concept to the Earth, Kelvin realised that it would be possible to determine its age by calculating the amount of energy available from all possible sources and then, from measurements (or estimates) of the rate at which energy is dissipated, determine the time necessary for the Earth to cool from its original molten state.

Despite the flimsy basis for many of the assumptions used in the calculations, Kelvin believed that the best measurements available were sufficiently exact to make meaningful calculations possible. The results he obtained indicated that the crust of the Earth was 98 million years (Ma) old, but that it would be prudent to place the acceptable limits at 20 and 400 Ma to take account of the uncertainties involved.

When these results were published in 1863, despite the resentment of many geologists at the intrusion of a non-geologist, the quantification proved highly attractive. It provided 'hard scientific' support to the long-held idea of a cooling, contracting, initially molten Earth, and because of this the associated 100 Ma timescale was also accepted. Thus by 1870 very few geologists defended a terrestrial age that differed much from this figure. In the first decade of this century, however, all of the assumptions that were fundamental to these ideas came under very severe attack.

The scientific basis for Kelvin's initially molten Earth was Laplace's nebular hypothesis, in which it was postulated that the Earth and all other planets had condensed out of a high-temperature rotating gas cloud. However, from about 1870 onwards, astrophysicists found more and more characteristics of the Solar System that could not be satisfactorily explained in these terms. This growing dissatisfaction led, in 1905, to the proposal that the Earth was the result of the low-temperature accumulation of much larger solid fragments (planetesimals), which appeared to resolve many of the problems raised by the nebular hypothesis.

An even more important prong of this attack came from the discovery by Curie and Laborde, in 1903, that the element radium maintains a temperature above that of its surroundings because of the spontaneous evolution of heat involved in its disintegration as a result of radioactivity. This was a source of energy Kelvin had been unable to foresee in 1863. The discovery of radioactivity led to a tremendous burst of research activity, headed by Ernest Rutherford. By early 1904 he was confident enough to state that the Earth could no longer be treated as a cooling body since radioactivity was constantly replenishing terrestrial heat by an unknown amount, and indeed temperature may have been sensibly constant for an extremely long time. This meant that Kelvin's calculations were probably of no significance, and furthermore there was no basis for the concept of a contracting Earth or of a restricted timescale.

Despite the severity of these attacks, geologists were by now very reluctant to abandon the idea of a timescale restricted to 100 Ma. Spurred on by Kelvin's intervention, they had made great efforts to develop purely geological methods for determining the age of the Earth, independent of the physicists. The most popular of these measured the present rates of erosion and/or sedimentation and, by comparing such figures with the total volume of sedimentary rocks, arrived at an estimate of the time since the processes of sedimentation started on the Earth's surface. A related but somewhat different method was developed in 1899 by John Joly, Professor of Geology at Dublin, which purported to give the age of the oceans. The basic ideas involved were simple: if it could be assumed that the oceans were initially condensed salt-free from the primordial atmosphere, that their store of sodium had been supplied at a near-uniform rate by rivers and that the sodium had remained in solution in the oceans since the process began, then the present amount of dissolved sodium divided by its rate of supply by the rivers would give the time required for its accumulation. In other words, the time since the Earth's surface cooled below 100°C would be known. The result obtained by these methods (80–90 Ma) appeared to be a magnificent vindication of the 100 Ma timescale.

The integrity and independence of such geological methods of age determination were not really questioned until 1913, when *The age of the Earth* by Arthur Holmes was published. This was the first full-scale review and assessment of the various methods of measuring geological time and was particularly significant in that it discussed the methods themselves, rather than dealing only with the numerical results (the normal procedure). In so doing Holmes exposed many inconsistencies that had been concealed by the apparent numerical agreement between the various methods.

However, he went further than this and completely vitiated the whole basis of age determinations based on geological processes. As stated previously, all of these methods assumed that the processes considered were operating at their present speeds in the past. While not denying the similarities in *kind* between past and present processes, Holmes argued that the same could not be said about their *rate* of operation, for at the present time the land areas of the Earth are more extensive and at a greater elevation than at other periods in the past. Hence contemporary rates of erosion and deposition must be that much greater and could not be used as a measure for rates throughout geological time.

In 1917 the same point was made even more forcefully by Barrell, Professor of Geology at Yale. He amassed great quantities of evidence to show that erosion and deposition are pulsatory processes dependent on the changing elevation of the continents and that this evidence had been largely ignored, or at best smoothed over, so that differences between past and present rates would appear minimal. Barrell agreed with Holmes that the present is a period of extraordinarily rapid denudation, perhaps 10–15 times greater than average, and furthermore that the relationships between denudation and sedimentation were so complex that there was no possibility of using them as a basis for age determinations and that an entirely independent method was required.

By 1920 the bases for Dana's model, together with the associated restricted timescale, were being very severely questioned. In spite of the seriousness of these attacks, most geologists were reluctant to abandon their old ideas. This was nowhere more apparent than in their reaction to a revolutionary idea which advocated that the continents had been moving great distances across the face of the Earth in the recent past, rather than being original and permanent features of the Earth's surface, and that such movements were still taking place. This idea was first presented as a short paper by Alfred Wegener in 1912 and in much more expanded form in the book *Die Entstehung der Kontinents und Ozeane* in 1915. Discussion on an international scale, however, did not occur until the third edition (1922) was translated into English with the title *The origin of continents and oceans*. Wegener was initially attracted to this concept by the way in which the continents on both sides of the Atlantic could, if moved together, be fitted closely against each other like the pieces of a jigsaw puzzle (Fig. 2.1). He then proceeded to marshal a great deal of evidence based on fossils and stratigraphy and the palaeoclimatic inferences that can be made from them.

In all these continents, now separated by thousands of kilometres of ocean, are found the fossil remains of the same group of trees and other

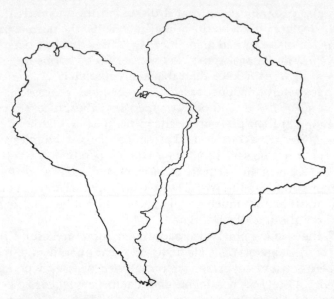

Figure 2.1 The jigsaw fit of continents (after Wegener).

plants that flourished in the Carboniferous period. They are typified by a seed-fern known, because of its tongue-like leaves, as *Glossopteris*. It was realised that such a distribution could not be explained either by the chance rafting of individual plants or seeds from one continent to another or by the simultaneous development of identical species at many widely separated sites. In 1885, Suess had suggested that the *Glossopteris* flora may have been established on a single supercontinent, Gondwanaland, great areas of which had subsequently foundered into the ocean basins, leaving the present continents as residual fragments. The distribution of the lemurs, the most primitive of primates and incapable of crossing any marine barriers, was explained in a similar way. They are now found on both sides of the Indian Ocean: to the west in Madagascar and closely adjacent areas of East Africa and to the east in India, Sri Lanka (Ceylon) and South-East Asia. Prior to Wegener's work, this was explained in terms of a supercontinent, Lemuria, most of which had subsided beneath the Indian Ocean.

It had also been recognised, from fossil and stratigraphic indicators, that climate at many places on the Earth's surface had changed with time. The abundant flora preserved in the Carboniferous Coal Measures of the Northern Hemisphere, and particularly the lack of growth rings in fossil trees, was taken to indicate an equatorial climate. Contemporary sediments in the southern continents consist of glacial deposits interbedded with *Glossopteris*-bearing sediments and coal beds, which were interpreted as resulting from a succession of glacial and interglacial periods.

This suggested that, since the Carboniferous, Europe's climate had

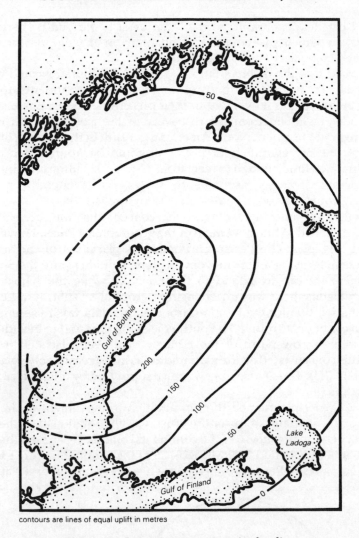

contours are lines of equal uplift in metres

Figure 2.2 Isostatic response to ice loading.

changed from tropical to temperate, while in the Southern Hemisphere the change had been in the reverse sense from polar to subtropical. Such climatic change was explained by many authors as being the result of movement of the Earth's geographical Poles relative to the crust, or vice versa.

Wegener recognised fundamental weaknesses in both of these arguments. With regard to the submergence of parts of former supercontinents, he saw that this was contrary to the principles of isostasy, for the granitic materials of the continents could never submerge in this manner. Gravity measurements showed that both continents and ocean basins were in a state of isostatic equilibrium, so the latter must be floored by very dense materials, in which case there was no room in the oceans for former continental materials of granitic composition.

It followed that the disjunct flora and fauna distributions could only be explained by the former contiguity of the present-day continents, followed by their break-up and movement to their present positions. Some support for this idea stemmed from the discovery of the isostatic uplift of both Scandinavia and northern North America as a result of the melting of the ice sheets of the last glaciation (Fig. 2.2). It could be argued that, as the substratum was fluid enough to react to such vertical loading and unloading of continents, it was reasonable to suppose that lateral continental movement could also occur in response to an applied force.

Generally, Wegener agreed with the concept that long-term climatic change was linked to the movement of the geographical Poles. However, he saw that to explain the Permo-Carboniferous glaciation of the southern continents in terms of a Pole movement alone would require the Southern Hemisphere ice cap to extend to the Equator, while the whole of the Northern Hemisphere was experiencing a tropical or subtropical climate. Wegener's solution for the Permo-Carboniferous data was to assemble the southern continents around the South Pole of the time and to include India, which, though now north of the Equator, had been glaciated from the south. This concentrated all the glaciated areas into a reasonably small polar ice cap (Fig. 2.3). This reassembly was also supported by other palaeoclimatic indicators (Fig. 2.4).

Thus Wegener maintained that, in addition to movement of the Poles, which he saw as a shift of the entire crust over the rotation axis, there was also independent movement of the continental blocks. He argued that, until the Carboniferous period (300 Ma ago), all of the continents were united in a single land mass – Pangaea. Then in a series of ruptures it broke apart, so

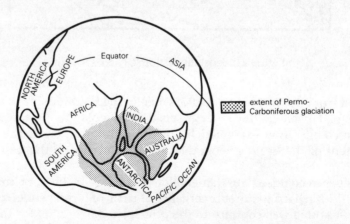

extent of Permo-
Carboniferous glaciation

Figure 2.3 Continental reassembly to explain evidence of Permo-Carboniferous glaciation of southern continents.

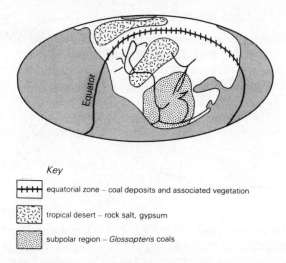

Key

┼┼┼┼ equatorial zone – coal deposits and associated vegetation

▨ tropical desert – rock salt, gypsum

▨ subpolar region – *Glossopteris* coals

Figure 2.4 Permo-Carboniferous climatic zones.

that during the Jurassic and Cretaceous periods (180–70 Ma ago) the southern continents, including India, separated from one another, while the final opening of the North Atlantic only occurred about 0.1 Ma ago, at the beginning of the Ice Age (Fig. 2.5).

The Mid-Atlantic Ridge together with Iceland and the Azores were regarded as continental material left behind when the continents flanking the Atlantic broke apart. The West Indies, Japan and the island arcs of the western Pacific were also thought of as continental material left behind in the westward drift of the continents. The mountains of the west coast of the Americas were the result of pressure at the leading edge of the westward-moving continental blocks, while the mountains of New Guinea resulted from the northward movement of Australia and the Himalayas from the northward pressure of India.

To explain the causative mechanisms of continental movement, Wegener suggested that two forces were involved: one responsible for movement towards the Equator and the other for movement towards the west (see box). While realising that both of these forces were extremely small, he believed that they would be effective if applied over a long period of time.

Wegener was convinced that his concept of continental movement could be tested by direct measurement of the longitude of a particular site over a period of years. Since he believed that the North Atlantic had opened in the last 0.1 Ma, where better than Greenland to determine such movement, particularly as longitudinal measurements were available for various sites in Greenland for the years 1823, 1870 and 1904. Calculations based on these figures showed that Greenland appeared to have moved westwards, relative to Greenwich, 420 m between 1823 and 1870 (9 m a^{-1}) and 1190 m between 1870 and 1904 (32 m a^{-1}).

Eotvos ('pole-fleeing') force, due to the oblate spheroidal shape of the Earth

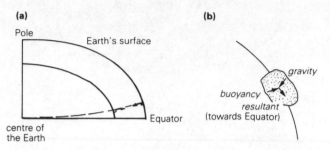

(a)

Pole

Earth's surface

Equator

centre of the Earth

(b)

gravity

buoyancy

resultant (towards Equator)

Except at the Poles and the Equator, a line drawn normal to the Earth's surface will not pass through the centre of the spheroid. Because the amount of flattening decreases with depth, a series of normals drawn at successively deeper levels describes a curve (a). Forces of buoyancy and gravity act perpendicular to the spheroid surface. Consider a 'floating' continent: gravity acts at the centre of gravity of the continent and pulls it perpendicularly downwards; buoyancy acts at the centre of buoyancy, which is the centre of gravity of the displaced medium in which the continent 'floats' (and so is lower than the centre of gravity of the continent), and pushes the continent perpendicularly upwards. Since these two forces are not exactly parallel, there is a small resultant force which tends to move the continent equatorwards (b).

Tidal attraction of both Sun and Moon

This exerts a drag on the entire crust, causing it to lag westwards relative to the interior. Since the continental blocks stand high, they are more affected than the underlying medium, so they tend to lag behind, i.e. move relatively westwards.

[Prepared by F. Williams]

From Wegener's point of view, not only did continental movement solve a great many geological problems that had otherwise defied solution but mechanisms for such movement had been proposed and the reality of movement had been established. Unfortunately, most of the geological community did not regard it in this light. In fact, reaction in many cases bordered on the hysterical, and the result was rejection of all aspects of the theory by the great majority of geologists.

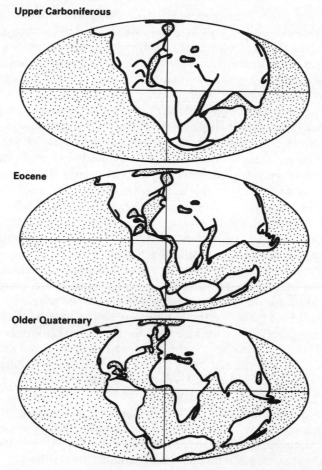

Figure 2.5 Three stages of continental displacement according to Wegener.

Fundamental to this rejection was that the whole idea of continental movement was revolutionary. It challenged opinions that had been held for so long and by so many geologists that they had become accepted truths and were not open to question. This was particularly so in the case of palaeontologists, who, rather than accept Wegener's much more elegant solution to the problems of animal and plant distribution, were prepared to continue with the geophysically impossible concept of sunken continental masses. In deference to this impossibility, however, they did reduce the transoceanic connections to very narrow land bridges; anything, as long as the continents remained fixed.

Complementary to this rejection was the argument that the fit of the continental margins was in fact very poor and that, to match the continents across the Atlantic, Wegener had considerably distorted the shapes of the

Americas and Africa. Another line of attack was to show that the coastline fit could be accidental: it was demonstrated that Australia could fit neatly into the Arabian Sea, which hardly coincided with Wegener's views. The ultimate counter-argument, however, was that a good continental fit could be used as evidence *against* movement, for the shape of the original break would have been entirely altered by erosion and a match would not be expected; so that, no matter what Wegener said, he would be wrong!

The attack of the geophysicists was much more to the point, for they were able to show that the proposed mechanisms were at least 10^6 times too weak to effect continental movement. They also pointed out a major contradiction that needed to be resolved. According to Wegener the continental crust could float in and drift through the mantle, so that the crust must be more rigid that the mantle. Yet Wegener also stated that mountain ranges were formed at the leading margins of moving continents, when the continental crust encountered the resistance of the mantle and was compressed into fold mountains, which meant that the mantle must be more rigid than the crust. The measurements that indicated movement of Greenland were also shown to be inaccurate. However, serious as these points were, there was no justification for the geophysicists' rejection of the whole concept. It is a classic example of throwing the baby out with the bath water.

For some geologists a reason for the rejection was that Wegener was not a geologist but a meteorologist, and he could be represented as an amateur dealing in things he did not really understand. This type of argument was used particularly by palaeontologists and stratigraphers to refute a great number of the suggested correlations.

For this combination of reasons, Wegener's ideas managed to affront practically the whole of the geological community. This was particularly so in the USA where, in 1926, under the guise of a scientific conference, they managed to stage something like a 'state trial'. This resulted in the personal ridiculing of Wegener as a scientist and the complete rejection of his ideas by the great majority of the participants.

However, there were a few people of sufficient stature who, while seeing the weaknesses, were still convinced of the fundamental merit of Wegener's thesis. Foremost among these were Arthur Holmes and A. L. du Toit, Professor of Geology from Johannesburg. Du Toit's own book on the subject, *Our wandering continents* published in 1937, provided a large number of additional examples of geological continuity across present continental boundaries, enabling him to reconstruct former supercontinents with a great deal more accuracy that had Wegener. This work led him to disagree somewhat with Wegener, for he saw evidence that initially there were two supercontinents rather than a single Pangaea: Laurasia in the north, consisting of parts of Europe, Asia, North America and Greenland, and Gondwanaland in the south, made up of parts of South America, India, Australia, Africa and Antarctica.

Du Toit also recognised the weakness of Wegener's postulated

(a) Peripheral to Laurasia

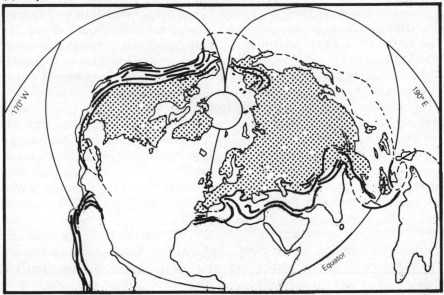

(b) Peripheral to Gondwanaland

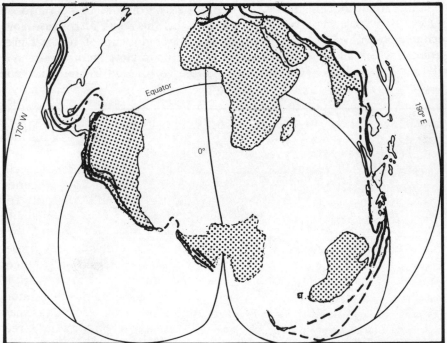

Figure 2.6 Position of fold mountains.

mechanisms of movement. He suggested an alternative that was an ingenious amalgam of Dana's idea regarding the formation of fold mountains at the periphery of continents and periodic subcontinental melting due to radioactive heating. He considered that the accumulation of sediments in subsiding marginal basins caused the continents to tilt and hence to have a tendency to slide gravitationally into the basins. This in turn would cause tensional effects and ultimately fractures within the continental mass.

This process would be aided by radiogenic heat, for the low thermal conductivity of the continental masses would prevent the escape of such heat from the underlying mantle. The resulting temperature rise would cause melting at the crust/mantle boundary. This combination of processes was seen to be responsible for the initial fracturing of Laurasia and Gondwanaland. The subsequent lateral movement of the fragments, while creating ocean basins in their wake, was also forcing up fold mountains from the basin sediments in front of them (Fig. 2.6).

Du Toit's work, concentrating as it does on evidence from the southern continents, meant little to the great majority of geologists, who had never been out of the Northern Hemisphere, and it was rejected as being merely a repetition of Wegener's ideas.

Within 40 years of its apparent rejection by much of the geological community, however, the main plank of Wegener's concept, the long-distance lateral movement of continents, had become accepted by most geologists. Such a remarkable turnaround was the result of considerable technological developments leading to the opening up of ocean basin geology, which allowed land-oriented geologists to view their subject in a totally different light. These themes will be considered in the next two chapters.

CHAPTER THREE

The development of geological technology

Geology in the 20th century has been marked by an ever-increasing application of physics and chemistry to the study of the Earth. Of particular significance have been the developments of radiometric dating, rock magnetism and seismology.

RADIOMETRIC DATING

Rutherford established in 1901 that, in the process of emitting α-, β- and γ-rays and producing heat, radioactive materials break down at a fixed rate. This occurs no matter what their temperature or pressure, or their state of chemical combination. He further realised that this property of regular breakdown could be used to determine the time at which the rocks and minerals containing radiocative materials became closed systems. The particular importance of this suggestion was that it provided a dating method that was completely independent of the rate at which geological processes operate, which was certainly not the case for the methods used in the last few decades of the 19th century.

Initially even the few geologists directly interested in the problems of absolute dating did not realise its potential. It seemed to have an even flimsier basis than the recently refuted method used by Kelvin and they regarded their own methods, such as Joly's, as being founded on hard factual data.

The majority of geologists were quite disinterested in these developments. They were preoccupied with correlation problems, the solution of which was mainly dependent on the fossil content of rocks. This had been one of the major centres of interest of 19th-century geology and had resulted in the development of a relative timescale, which they felt served their purposes admirably. It mattered little to them whether 100 or 500 Ma had passed since the first fossiliferous rocks were formed.

Before radiometric dating could become generally acceptable, therefore, its credibility had to be established on physical grounds. It also had to be shown to be applicable to a wide range of rocks, both in composition and geographical distribution, so that it could be fused with the relative timescale to produce a unified chronological scheme. Before following these developments, however, the relative timescale (Table 3.1) needs to be discussed to make its nomenclature rather more understandable.

Table 3.1
Relative timescale based on fossil contents of rocks.

Era	Decreasing periods of time ⟶	
	Period	Epoch
Cenozoic	Quaternary	Holocene
		Pleistocene
		Pliocene
		Miocene
	Tertiary	Oligocene
		Eocene
		Palaeocene
Mesozoic	Cretaceous	
	Jurassic	
	Triassic	
Palaeozoic	Permian	
	Carboniferous	
	Devonian	
	Silurian	
	Ordovician	
	Cambrian	

One of the major problems of 18th-century geology was the correlation of one sequence of sedimentary rocks with another. Initially, general characters were used, such as whether the rocks were sandstones, limestones or shales, or whether they were strongly folded or not. This was reasonably satisfactory over short distances but, because many rock units tend to change in character laterally, reliability decreased markedly with increasing distance. This state of affairs continued until it was realised, by William Smith in England and Cuvier and Brongniart in France, that fossils were a much more reliable guide to correlation. They showed that each formation of sedimentary rocks contained a unique suite of fossils, which never occurred in either earlier or later rocks.

In the 19th century, geology was preoccupied with working out this idea and applying it progressively to sedimentary formations that were more difficult to interpret and to the remaining areas on all continents. By the end of the 19th century this had resulted in the development of a relative timescale based on fossil content which had world-wide applicability, covering all sedimentary rock formations from Cambrian times onwards.

The nomenclature of the relative timescale was derived from a number of sources; some names indicated the type of fossils, others derived from the locality where particular rocks were first studied and yet others were inherited from earlier times when only general rock characters were used for correlation.

At the highest level, three **eras** were distinguished according to the dominance of particular fossil groups. In the oldest of these, marine invertebrates, termed ancient life forms, were dominant and hence it was called the **Palaeozoic**. This was succeeded by the era of middle life forms, or **Mesozoic**, in which dinosaurs were dominant, together with marine invertebrates, such as the ammonites. The third was the era of recent life forms, or **Cenozoic**, which was dominated by mammals, birds, insects and angiosperm plants. These three terms were proposed around 1840 and replaced others based on general rock characters. Vestiges of these are still present today in the names **Tertiary** and **Quaternary**, used for the two most recent **periods**. These names originate from a scheme developed in the 18th century, in which rocks were classified as being primary (crystalline), secondary (consolidated) or tertiary (alluvium). Only the last of these has survived, although it is now applied to a much older sequence of materials than the original alluvium. Alluvium is now included in the Quaternary period, a term in keeping with the other three but not coined until the 19th century.

Three periods make up the Mesozoic and all were named by European geologists according to general rock characteristics. Only later did they come to have a fossil connotation. The **Cretaceous** was named in the 1820s after the chalk that is such a feature of the coasts of the eastern English Channel. The **Jurassic** was named in 1799 by the great naturalist Alexandre von Humboldt after a series of massive limestones in the Jura Mountains. Unfortunately, most of these rocks are now classified as Triassic rather than Jurassic, but even so the use of the name has persisted. The name **Triassic** was first applied in the 1830s to rocks occurring in northern Germany. The name derives from the fact that these rocks consisted of three stratigraphic units, in upward sequence sandstones, limestones and marls.

In contrast, all of the Palaeozoic periods, apart from the Carboniferous, were named by British geologists after particular localities and all on the basis of their fossil content. These six periods may be split into two groups of three: the Permian, Carboniferous and Devonian of the Upper Palaeozoic, and the Silurian, Ordovician and Cambrian of the Lower Palaeozoic. The names of the latter group arose because these rocks were first investigated between North Wales and the English border. Adam Sedgwick, Professor of Geology at Cambridge, started in North Wales from the bottom of the sequence and worked upwards. Meanwhile Murchison, later

Director of the Geological Survey, started in the Welsh borderlands from the top of the sequence and worked downwards. Sedgwick referred to his rocks as **Cambrian**, after Cambria, the Roman name for Wales, while Murchison called his rocks **Silurian**, after a Celtic tribe that had inhabited this part of Britain. Because there was no agreement as to where the two systems should meet, the end result was a massive overlap between the two and a great deal of acrimony between the geologists involved. This situation continued for about 40 years until the name **Ordovician** was suggested for the disputed intermediate strata. This name was also derived from a Celtic tribe that had inhabited the area.

The **Carboniferous** rocks in the middle of the Upper Palaeozoic were recognised throughout western Europe at an early date because of the economic importance of their coal-bearing strata, after which they were named. Sedgwick and Murchison were once again involved in naming the earlier **Devonian** period. In the late 1830s, after much of their fieldwork in Wales had been finished, but before the disagreement about results had developed, they began work on the folded and crushed rocks that underlie the Carboniferous in Devon. This was on the supposition that they were equivalent to Sedgwick's North Wales Cambrian. However, this correlation based on general rock character was refuted when fossils showed that the strata were intermediate between the Silurian and the Carboniferous, so a new period was recognised and called Devonian because of its geographical location.

In western Europe the rocks overlying the Carboniferous were generally poor in fossils, so when Murchison visited Russia in the 1840s and found that equivalent rocks were richly fossiliferous, he proposed that this last period of the Palaeozoic be called the **Permian**, after the locality west of the Urals where these rocks occur.

Charles Lyell, another major figure of 19th-century geology, was the first to recognise **epochs** within the Tertiary and Quaternary. He realised that sediments of these periods occurring in southern England, France and Italy contained a decreasing percentage of extant species of shells with increasing age, which enabled him to differentiate three epochs that, from oldest to youngest, were named the **Eocene** (*Eos* = dawn and *cene* = recent), **Miocene** (*Mio* = less than) and the **Pliocene** (*Plio* = more than). Subsequently it was realised that there was a major break in the Pliocene, so that it was split into an Older Pliocene and Newer Pliocene. As this proved to be somewhat confusing, the Newer Pliocene was renamed **Pleistocene** (*Pleisto* = the most, i.e. the most like the present) and the Older Pliocene reverted to simply Pliocene.

All these nomenclatural innovations occurred within the 1830s and were based on a small number of locations. Subsequent fieldwork in Germany and France showed that there were gaps in Lyell's sequence on both sides of the Eocene. Lyell's style of nomenclature was preserved, however, by naming the time range of the newly discovered strata above the Eocene as

Table 3.2
Typical figures illustrating radioactive decay.

successive equal time periods	1	2	3	4	5	6
amount present at the beginning of that time period, N (g)	12.00	9.00	6.75	5.06	3.80	2.85
amount that decays in that time period n (g)	3.00	2.25	1.69	1.26	0.95	0.71

the **Oligocene** (*Oligo* = few) and that below the Eocene as the **Palaeocene** (*Palaeo* = old).

There were yet further developments when it was recognised that the glacial deposits of northern Europe were part of the Pleistocene and as such were markedly different from the succeeding Postglacial deposits, which were referred to as **Holocene** (*Holo* = complete) as they contained a fauna identical with that in existence at present. As there is such a profound faunal break at the base of the Pleistocene, the Pleistocene and Holocene are considered together as being the two epochs of the Quaternary period, the most recent part of the Cenozoic era.

This timescale is relative in that the order of periods is clearly established, although the actual age of the rocks or fossils is unknown. Rutherford's investigations into radioactivity promised to change all this. It was realised that radioactive material decays in a regular manner to give a stable end or **daughter** product. It followed that, if the rate of decay was known and the amount of radioactive **parent** and daughter present in a rock (or mineral) could be measured, then it would be possible to calculate the time since the rock had formed, provided that no radioactive parent and no daughter products had been added or removed in this period of time.

The crux of understanding radiometric dating lies in comprehending the nature of radiometric decay, and so this needs to be examined in somewhat greater detail. The typical set of figures (Table 3.2) gives both the amount of a particular radioactive material present at the beginning of each of six successive equal timespans and the amount of the radioactive material that decayed during each period. From this it can be seen that n/N is a constant (in this case 0.25) referred to as λ, which is characteristic of all radioactive decay. This is also reflected in the curve produced by plotting N against time (Fig. 3.1), which gets ever nearer to zero with increasing numbers of time periods, but never quite makes it. All radioactive decay follows such an exponential curve, although the constant λ (the rate of decay) differs from one radioactive parent to another. The shapes of decay curves vary with the value of λ from that given in Table 3.2 where $\lambda = 0.25$, a very high value, to an almost horizontal line when λ is very small (Fig. 3.1). Nevertheless, all decay curves are exponential, and from knowing one point on such a curve it is possible to calculate any other point on it by using natural logarithms. This is fundamental to solving the problem of radiometric dating.

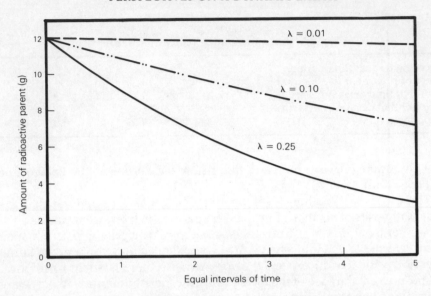

Figure 3.1 Radiometric decay curves.

Figure 3.2 shows the exponential decay curve of a particular radiometric parent. The amount of parent now present in a sample is known (N_p) as well as its decay constant (λ). The amount present at any time in the past, say at points 1 and 2, is then given by the expressions $N_p e^{\lambda t_1}$ and $N_p e^{\lambda t_2}$ respectively (where 'e' is the base of natural logarithms).

It is possible using these expressions to determine t, the time involved, which is of course the unknown in radiometric dating, for it can be seen that, where N_d is the end or daughter product,

$$N_p e^{\lambda t_2} - N_p = N_{d2}$$

and

$$N_p e^{\lambda t_1} - N_p = N_{d1}$$

In more general terms, this means that

amount of radioactive parent present at a given time in the past	−	amount of radioactive parent now present	=	amount of end or daughter product

or

$$N_p e^{\lambda t} - N_p = N_d$$

Analysis of the particular sample of rock or mineral to be dated will give N_p

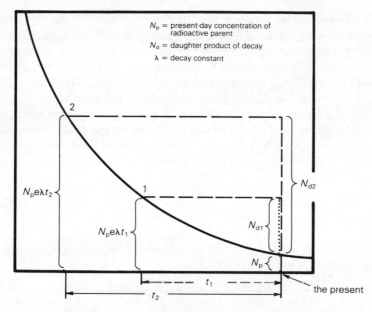

Figure 3.2 Decay curve and radiometric dating.

and N_d and λ is known from experimental results. As t is the only unknown in the equation, it is determinable. The equation can be rearranged so as to isolate t (see box) and arrive at the following equation

$$t = (2.303/\lambda) \log_{10}(1+N_d/N_p)$$

We wish to find an expression for t from the equation:

$$N_p e^{\lambda t} - N_p = N_d$$

Dividing by N_p and transposing, gives

$$e^{\lambda t} = 1 + N_d/N_p$$

Taking natural logarithms on both sides:

$$\lambda t = \log_e (1 + N_d/N_p)$$

and transposing:

$$t = \frac{1}{\lambda} \log_e (1 + N_d/N_p)$$

Changing to ordinary logarithms then gives

$$t = (2.303/\lambda)\log_{10}(1 + N_d/N_p)$$

This equation needs another slight modification. The only radioactive parents useful for geological dating decay very slowly, for otherwise after a period of geological time there would be nothing left to measure. This makes λ, normally expressed as the amount of decay per year, extremely small and rather difficult to compare from one radioactive parent to another. A more convenient way is to express radioactive decay in terms of the time required for half a given quantity of a particular radioactive parent to decay. This is naturally called the **half-life** (T) and is generally expressed in millions of years (Ma). It is just as much a constant for each radioactive parent as is λ, to which it is inversely related by the expression

$$T = {}^{0.693}/\lambda$$

Substituting

$$\lambda = {}^{0.693}/T$$

in the last equation gives

$$t = 3.323T \log_{10} (1 + N_d/N_p)$$

It was on this basis that the first radiometric determinations were made on rare uranium-rich minerals, in which the parent element was uranium and the decay products helium or lead. The results gave ages that ranged from 410 to 2200 Ma and in view of the likely loss of helium from the minerals they could only be regarded as minimum values. These great ages, although put forward tentatively, provoked strong reactions from most geologists, including John Joly, who still supported 100 Ma as the age of the Earth.

It was argued that the geological methods of age determination, particularly that based on the sodium content of sea water, were much the best and most exact methods available and, as the results of radiometric dating were different, there must be something fundamentally wrong with the radiometric method. As seen in Chapter 2, the work of Holmes and Barrell on rates of erosion was sufficient to demonstrate the weakness of these arguments.

However, Joly was not content just to say that there must be something wrong with the radiometric method, he set out to prove it. He realised that radiometric dating was based on the supposition that the radioactive decay constant (λ) had never varied over geological time. However, the evidence for the constancy of breakdown was no more than circumstantial. A possibility of testing this arose from the discovery that, in thin sections of certain minerals, such as biotites in granites, there were small circular dark spots or concentric rings called **pleochroic haloes**. These were 12–42 μm in diameter and occurred around tiny inclusions of other minerals, such as sphene and zircon. It was established that these haloes were produced by α-particles shot out during the decay of radioactive parent elements concentrated in the sphene and zircon. In addition it was known that for a given host mineral and a given radioactive parent, the distance travelled by such α-particles is proportional to the decay constant (λ). Therefore the size of the haloes from minerals of different ages could be used as a measure of the possible variation of λ with time. Evidence was produced that uranium haloes were larger in Precambrian rocks than in more recent rocks, pointing to a decrease in λ with time.

Such difficulties were resolved as research gradually clarified the whole process of radioactive decay. It was established that the nucleus of every atom, apart from that of hydrogen, consisted not only of positively charged **protons** but also **neutrons**, particles with practically the same mass as protons but no charge. In these terms an element consists of atoms with the same number of protons, but atoms of a particular element could differ from one another in terms of their mass by containing different numbers of neutrons, and such atoms were referred to as **isotopes**. Furthermore, it was recognised that the process of radioactive decay was dependent on certain isotopes with unstable nuclei and these ejected particles until a final stable state was reached. Therefore, to achieve accurate results from radiometric dating, it was necessary to determine the amounts of both parent and daughter products (N_p and N_d of the basic equation) in terms of isotopes, rather than as elements, which was all that had been possible using chemical methods of analysis. Such determinations only became possible with the development of the mass spectrometer, which separated atoms from one another according to their mass.

By 1930 with the use of this instrument it had been shown that there were two isotopes of uranium (with masses of 238 and 235) and one of thorium (mass 232) involved in natural radioactive decay and that each of these isotopes had very different half-lives (Fig. 3.3). In the process of decay, ^{238}U lost eight particles each consisting of two protons and two neutrons (the α-radiation of Rutherford). These particles then formed helium and left behind a daughter or stable isotope of the element lead with a mass of 206, ^{206}Pb. In the same way, ^{235}U lost seven such particles to give rise to another

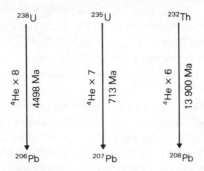

Figure 3.3 Isotopic decay, showing parents, their half-lives and end products.

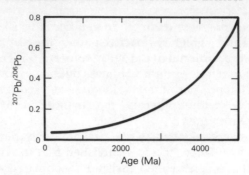

Figure 3.4 ^{207}Pb/^{206}Pb age determination.

lead isotope, ^{207}Pb, and ^{232}Th lost six particles to produce yet another lead isotope, ^{208}Pb.

This meant that radiometric age could be determined by evaluating the ratio between these three parent isotopes and their daughter products, i.e. ^{238}U/^{206}Pb, ^{235}U/^{207}Pb and ^{232}Th/^{208}Pb. Use could also be made of the ratio between ^{207}Pb and ^{206}Pb, for as ^{235}U (the parent of ^{207}Pb) decays at a much faster rate than ^{238}U (the parent of ^{206}Pb) it follows that with increasing age a mineral containing uranium would have an increasing ratio of ^{207}Pb/^{206}Pb (Fig. 3.4). If, in a particular mineral or rock, the age given by any one of these methods was not consistent with any of the others, the results could not be considered reliable.

Reassessment in these terms by Arthur Holmes in the early 1930s found that very few past determinations were acceptable; in fact the total number was seven. This greater understanding of the radioactive process also seemed to resolve the controversy over the size of pleochroic haloes, for it was then possible to explain the large haloes in Precambrian rocks in terms of the faster decay of ^{235}U and the smaller haloes as a result of the slower decay of ^{238}U. Even though doubts have subsequently been raised about this interpretation, it was sufficient to refute Joly's contention that λ varied with time.

Therefore, by the 1930s, reliable and generally acceptable methods of radiometric dating had been developed, although these were only applicable to rare, highly radioactive minerals. Even then the number of minerals that gave consistent results was very much smaller again. The way forward from this position depended upon the development of techniques whereby smaller amounts of radioactive isotopes and their decay products could be accurately analysed.

It was known that uranium and thorium and their related lead isotopes occur in extremely small amounts in minerals such as zircon, sphene and apatite, which are widely distributed as minor constituents of granites. Thus, if analyses of the required accuracy could be achieved, the uranium/thorium/lead dating methods would have a much wider applicability.

It was also known that radioactive decay occurred in both rubidium and potassium. The radioactive rubidium isotope is ^{87}Rb, which decays by nuclear transformation (a neutron becoming a proton) and ejection of an electron (Rutherford's β-radiation). This means that the mass does not alter but the number of protons increases by one, so that instead of being an atom of rubidium with 37 protons it becomes an atom of strontium with 38 protons. The reaction can be represented as follows:

$$^{87}_{37}\text{Rb} - \text{electron} = ^{87}_{38}\text{Sr}$$

where the lower figures, 37 and 38, indicate the number of protons (the **atomic number**) and the upper figure of 87 the sum of protons and neutrons (the **mass number**).

The breakdown of potassium is more complex. The radioactive isotope is ^{40}K and 91.5% of it behaves like ^{87}Rb, so that the potassium atom with 19 protons changes into a calcium atom with 20 protons:

$$^{40}_{19}\text{K} - \text{electron} = ^{40}_{20}\text{Ca}$$

However, since calcium is a very common element and ^{40}Ca is its commonest isotope, there is little possibility of distinguishing the ^{40}Ca resulting from potassium decay, so generally it is of no value for dating purposes. In contrast, the remaining 8.5% of the ^{40}K changes by the nucleus capturing an orbital electron, so that a proton becomes a neutron and an atom of ^{40}Ar is produced:

$$^{40}_{19}\text{K} + \text{electron} = ^{40}_{18}\text{Ar}$$

The inertness of argon and the rarity of this particular isotope makes this reaction particularly suitable for dating.

As both rubidium and potassium are widely distributed in rocks and minerals, dating by these decay schemes offered tremendous possibilities. However, before this could be achieved, techniques had to be developed for detecting extremely small amounts of both the parent and daughter products of these reactions. In the case of rubidium, this was because it is a rare element that occurs only as a minor component in potassium minerals and because the radioactive isotope ^{87}Rb only forms a quarter of this already very small amount. In addition the ^{87}Rb/^{87}Sr decay is very slow, with a half-life of 50 000 Ma (over 10 times greater than the half-life of ^{238}U), so that even in the best of rock or mineral samples the amount of both ^{87}Rb and ^{87}Sr is extremely small.

The element potassium is much more abundant than rubidium, but of its three isotopes the radioactive ^{40}K amounts to only 0.000 119%. Out of this only 8.5% decays to produce ^{40}Ar, the only product useful for dating purposes. Again the half-life of this decay is fairly large (11 850 Ma), being more than twice that of ^{238}U. In this case the complexity of the decay scheme, the small absolute amounts involved and the necessity of being able to handle and measure very small volumes of argon gas presented immense technical problems.

New techniques were developed in the 1940s for separation of the minute quantities of isotopes involved. When these were combined with the technique of isotope dilution (see box), the last of the major hurdles was cleared, for this enabled these extremely small amounts to be determined to the accuracy required for dating purposes.

Isotope dilution

The technique of isotope dilution depends on the fact that the *ratio* between two isotopes can be more accurately determined, in a mass spectrometer, than can the *absolute amounts* of either one of them. In the case of $^{40}K/^{40}Ar$ determinations, isotope dilution involves mixing the small but unknown amount of the daughter ^{40}Ar (extracted from a given weight of the mineral to be dated) with an equally small but accurately known amount of a stable argon isotope, generally ^{38}Ar. The ratio between the two mixed isotopes is then obtained by means of the mass spectrometer. Knowing the absolute amount of ^{38}Ar added to the mixture, the absolute amount of the ^{40}Ar can be calculated. Supplies of the purified isotopes necessary for this technique became generally available after 1945.

In the case of the other dating methods, additional problems arose when determining the extremely small amounts of radioactive isotopes and decay products. The uranium/lead method was first applied only to rare, highly radioactive minerals where relatively large amounts of ^{238}U, ^{235}U, ^{206}Pb and ^{207}Pb were present. Any lead initially present when the crystals formed was so small in comparison that it could be safely ignored. When zircon, sphene and apatite were being dealt with, however, the amounts of the lead isotopes being assessed were much smaller and so it became absolutely essential to differentiate between radiogenic and inherited isotopes.

This problem was solved by the development of isochron diagrams (see box), which allowed the full potential of the new technology to be applied not only to the straightforward dating of rocks but also to the solving of complex problems of Earth history. Nothing illustrates this better than the problem of the age of the Earth.

The solution depends upon accepting meteorites (see Ch. 1) as collections of minerals that have been radiogenically closed systems since the time of planetary formation, early in the history of the Solar System. On analysis, iron meteorites were found to contain such extremely small

Isochron diagrams

If four minerals were taken for analysis from an igneous rock shortly after its formation, they would be found to contain different amounts of uranium and lead because of differences in the chemistry of these two elements. However, the isotopic ratios of the uranium and the lead would be the same in each of the four minerals. This means that the ratio between any two lead isotopes, say ^{206}Pb (from previous ^{238}U decay) and ^{204}Pb (a non-radiogenic isotope), would be the same in all four minerals, while that between a uranium and a lead isotope, say ^{238}U and ^{204}Pb, would differ in each case according to the chemistry of the two elements. Plotting these ratios against one another (a) would thus produce a straight line ABCD at right angles to the ordinate because the only variable is the uranium content, which increases from mineral A to mineral D.

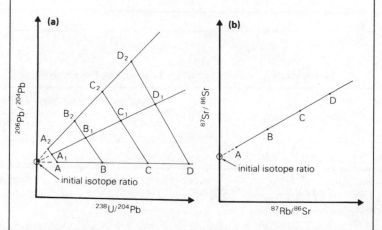

If these same analyses were repeated after the lapse of a significant amount of time, a certain amount of ^{238}U in all four minerals would have decayed to ^{206}Pb, while the ^{204}Pb would have remained the same. In terms of the ratios, ^{206}Pb/^{204}Pb would have increased, while ^{238}U/^{204}Pb would have decreased, but the effect would have been greatest in the case of mineral D, for this initially contained the greatest amount of ^{238}U, and least for mineral A, which initially contained the least. Plotting these results would give another straight line $A_1B_1C_1D_1$ which, if extended, would cut the ordinate at the initial ^{206}Pb/^{204}Pb ratio. Therefore plotting these isotope ratios for a number of minerals from any given igneous rock will enable the initial ^{206}Pb/^{204}Pb ratio to be determined. This will then allow the basic decay equation

$$t = 3.323\,T \log_{10} (1 + N_d/N_p)$$

to be applied to each of the four minerals, for now it is possible to differentiate between the ^{206}Pb produced by radiogenic decay since the rock crystallised (which is N_d in the equation) and the rest of the ^{206}Pb that was inherited. All four minerals should give the same age, for all were formed at the same time and have been subject to ^{238}U/^{206}Pb decay for the same period. Therefore, all straight lines of this type are lines of equal time, or **isochrons**, and their gradient is indicative of that age. The older the rock the steeper is the gradient, for it can be seen that after a further period of time the ratios of the isotopes in each of the minerals would plot as A_2, B_2, C_2 and D_2 and the straight line connecting these points, the isochron, would have a steeper gradient.

These isochron diagrams were applied not only to uranium/lead but also to rubidium/strontium dating. In this case the parent ^{87}Rb and daughter ^{87}Sr isotopes are compared to the non-radiogenic isotope ^{86}Sr (b) to arrive at a measure of the initial ^{87}Sr/^{86}Sr ratio.

Table 3.3
Isotopic composition of lead and the age of the Earth

	Relative isotope composition		
	^{204}Pb	^{206}Pb	^{207}Pb
Meteorites			
(a) Irons			
(i) Henbury, Australia	1	9.55	10.38
(ii) Canyon Diablo, Arizona, USA	1	9.46	10.34
(b) Stones			
(i) Forest City, Iowa, USA	1	19.27	15.95
(ii) Modoc, Kansas, USA	1	19.48	15.76
(iii) Nuevo Laredo, Mexico	1	50.28	34.86
Present-day lead			
(a) red clay	1	18.95	15.76
(b) manganese nodules	1	18.91	15.69

amounts of uranium and thorium (Table 3.3) that it could be assumed that the lead present was in no way radiogenically derived and hence must have been primaevally inherited. As such, this was lead that had remained virtually unchanged since the time of planetary formation. The higher values of uranium and thorium found in stony meteorites meant that their lead was a mixture from both primaeval and radiogenic sources. As the primaeval lead contribution was known, from the iron meteorite data, the radiogenic contribution could be determined. From these results the ratio of radiogenic $^{207}Pb/^{206}Pb$ was found to be 0.57, which gives an age for the stony meteorites of 4500 Ma (Fig. 3.4). It can be assumed that the iron meteorites are of the same age as the two types of stony meteorites, so that by plotting $^{207}Pb/^{204}Pb$ against $^{206}Pb/^{204}Pb$, three points will be obtained on an isochron for 4500 Ma (Fig. 3.5).

As this can be taken as an indication of the time of general planetary formation, it is probable that the Earth was formed at about the same time. The problem then was how to obtain hard evidence to confirm this probability. The closed systems of the meteorites are thought to be derived from materials equivalent to the mantle and the core of the Earth. As these zones have not been penetrated, no samples of such materials are available on the Earth. The material that is available from the lithosphere is being profoundly affected by so many processes that no rocks can be found that have maintained a closed system for anywhere near this length of time. The solution lay in considering the Earth as a whole as a closed system. In these terms since the Earth was formed, that is became a closed system, lead from radiogenic sources has been continuously added to the original stock of primaeval lead, so that if a good average sample of the Earth's lead at the present-day could be obtained it would be equivalent to the lead of a stony meteorite. The red clay and manganese nodules that are such a feature of large areas of the Pacific Ocean floor contain such a sample of lead, for it has been derived from a wide variety of sources distributed all round the Pacific Basin and thoroughly mixed in the oceanic waters, prior to its deposition

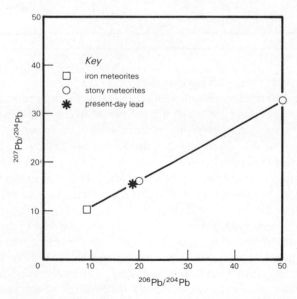

Figure 3.5 Lead/lead isochron and the age of the Earth.

and incorporation in these materials. Plotting the ratios of $^{207}Pb/^{204}Pb$ against $^{206}Pb/^{204}Pb$, resulting from the analyses of these nodules and clay, located a point on the 4500 Ma isochron of Figure 3.5. Therefore, radiogenic lead has been developing on the Earth for 4500 Ma. That is, the Earth was formed 4500 Ma ago at the same time as the meteorites.

Table 3.4
Unified timescale

Era	Period	Epoch	Radiometric age (Ma)
	Quaternary	Holocene	0–2 or 3
		Pleistocene	
Cenozoic		Pliocene	2 or 3–12
		Miocene	12–25
	Tertiary	Oligocene	25–40
		Eocene	40–60
		Palaeocene	60–70
	Cretaceous		70–135
Mesozoic	Jurassic		135–180
	Triassic		180–225
	Permian		225–270
	Carboniferous		270–350
Palaeozoic	Devonian		350–400
	Silurian		400–440
	Ordovician		440–500
	Cambrian		500–600
Precambrian			600–4500

All these developments and particularly the very great expansion in the use of the K/Ar and Rb/Sr methods since 1950 have made it possible to agree upon a unified timescale covering the whole of geological time. This has given absolute ages to the geological periods which had been established in a sequence by the use of fossils (Table 3.4). The dating of the relative timescale was not easy, because radiometric methods are generally applicable only to igneous and metamorphic rocks, while the relative timescale was determined by fossils in sedimentary rocks. However, there are enough situations in which the different types of rocks are interfingered to give a reasonable estimate of age in years to fossil-bearing rocks (Fig. 3.6).

From this work emerged the realisation that the Precambrian period was six times as long as that part of geological time covered by the relative timescale: 3900 Ma compared to 600 Ma. The continuing application of the radiometric dating method is now slowly disentangling the history of the Earth during this immense period of time.

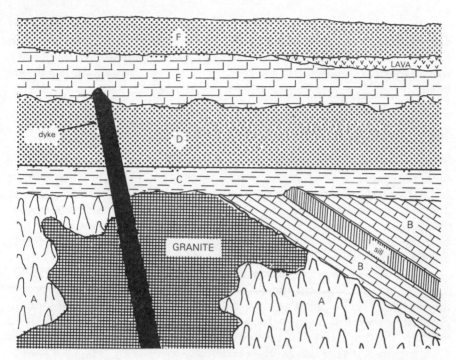

Figure 3.6 Stratigraphy, fossils and radiometric dating in geological chronology.

Relative timescale based on stratal relationships	Relative age established by means of fossil content of sedimentary rocks	Radiometric age of igneous rocks (Ma)
F	Pliocene	
erosive interval during which lava flows were erupted		lavas, 26±2
E	Eocene	
erosion		dyke, 130±5
dyke intruded		
D	Jurassic	
C	Triassic	
eroded		sill, 270±10
tilted		
intruded by sill		
B	Carboniferous	
interval during which granite was intruded and great erosion occurred		granite, 359±10
A	Silurian	

By the beginning of the 1960s, therefore, methods were available for dating a great range of Earth materials and a timescale had been constructed covering the whole of Earth's history with world-wide applicability. This was an essential tool for the development of a new Earth model in the late 1960s.

ROCK MAGNETISM

In the second half of the 19th century it became apparent that certain surface rocks possessed weak magnetic properties and that this magnetism resided in the iron-rich minerals that they contained. Progress in understanding this phenomenon was slow, for there were theoretical problems in understanding how the magnetism occurred and practical problems in devising instruments sensitive enough to measure such weak fields. It was not until the 1940s that some generalisations began to emerge.

By this time it was possible to say that some rocks acquire a stable magnetism at or about the time of their formation, which is a reflection of the Earth's magnetic field of that time. Unless disturbed by subsequent physical or chemical changes this **natural remanent magnetism** provides a record of the direction of the Earth's field at a particular spot, at a particular time in the past; that is to say, it can be used as a fossil compass. This natural remanent magnetism can be described in terms of **declination** and **inclination**, declination being the degree of deviation of a compass needle from geographical north and inclination its angle of dip below the horizontal (Fig. 3.7).

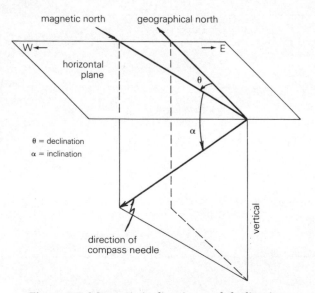

Figure 3.7 Magnetic inclination and declination.

Natural remanent magnetism was demonstrated experimentally and was also checked in the field by examining lava flows of known age at sites where magnetic records of declination and inclination were available, such as at Etna in Sicily, Hecla in Iceland and various Japanese volcanoes.

It was recognised that this magnetism could be acquired in several different ways, dependent upon the mode of formation and the subsequent history of the rock involved. Rocks formed by cooling from a high temperature acquire most of their magnetism as they pass through a particular temperature, the Curie point. This **thermoremanent magnetism** is preserved in such rocks as lava flows, dykes and sills and gives a reliable indication of the direction of the Earth's field at the time when the rocks cooled below the Curie point.

Sediments can also become magnetised by the preferred orientation of small grains of magnetite, generally no greater than 1–2 µm in size, when they are deposited with other detrital minerals in calm water in the presence of the Earth's magnetic field. The small grains of magnetite tend to align themselves with the Earth's field, giving a feeble but measurable **depositional remanent magnetism**. When these conditions of sedimentation were duplicated in the laboratory and the direction of magnetism was determined after the resulting deposit had dried out, the horizontal direction (declination) agreed closely with the Earth's field, but the vertical direction, or inclination, was much less. This was attributed to two effects: the tendency for any elongated grain (whether magnetic or not) to settle horizontally, and the subsequent compaction, which emphasised this tendency by rotating any elongated grains into an even more horizontal position. Thus sediments recovered from old lake beds and from deep-sea cores, while giving reliable measurements of declination, give inclination values considerably less than those of the Earth's field at the time the sediments were formed.

The extent to which depositional remanent magnetism survived the process of compaction and lithification (when sediments are transformed into rocks) was largely unknown. These processes are marked by changes in the sediments as materials in solution are precipitated within the pore spaces. In many cases this gives rise to a **chemical remanent magnetism**, as is the case in the red sandstones that are such a feature of Devonian and Triassic rocks in many parts of the world. The natural magnetism in this case is due to haematite, introduced in solution shortly after the sediment was deposited, which now both coats the individual quartz grains and acts as a cement between them. Little is known of how this magnetism is acquired at a temperature far below the Curie point, but it is widely used in palaeomagnetic determinations, for it is very stable and gives a good measure of both magnetic declination and inclination of the Earth's field at the time the haematite was precipitated.

It was well known that the direction of the Earth's magnetic field in terms of both declination and inclination is variable, even over a few years. The

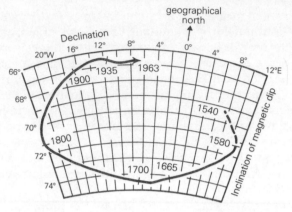

Figure 3.8 Secular variation in the Earth's magnetic field.

rapidity of these **secular variations** is illustrated by the records kept in London since 1546 (Fig. 3.8). In these circumstances it seemed that knowledge of the fossil field direction would not be of great significance. However, when viewed over several thousand years, instead of a few hundred, they assumed a totally different significance. The results of an investigation into the variation of declination given by clay sediments deposited over the period 15 000–9000 BC in New England, USA (Fig. 3.9), showed many variations on either side of true north. If the mean declination was calculated for this clay sediment over these several thousand years, it would approximate to the present direction of geographical north. It could then be assumed that over such a time interval the magnetic field Poles coincide with the geographical or rotational Poles. Theoretical support for this conclusion came from the idea that the form of the Earth's magnetic field is controlled by differential rotational movements at the core/mantle boundary, so that coincidence between the rotational and magnetic Poles is to be expected.

The natural remanent magnetism of the New England clay sediments was

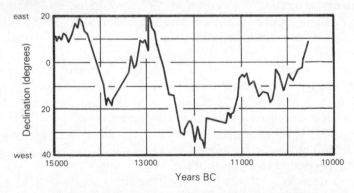

Figure 3.9 Declination of American varved clays.

depositional, which meant (as explained above) that, while declination values were a faithful copy of the Earth's magnetic field, the inclination values were probably too small. However, if mean values were determined for a series of lava flows (thermoremanent magnetism), both declination and inclination would accurately reflect the state of the Earth's magnetic field. Not only could the direction of true north be determined, but also the latitude of the lava flows, for the inclination of the Earth's magnetic field changes systematically with latitude such that

$$\tan I = 2 \tan L$$

where I is inclination and L is latitude. Thus an inclination of 60° indicates a latitude of 41°, for $\tan 60° = 1.7321 = 2 \tan 41°$.

In such cases, then, both the distance and the direction of the geographical Pole from a particular locality can be determined. It is therefore possible to locate the Pole position on the globe at the time the sampled rocks received their magnetism (Fig. 3.10). When such Poles were plotted on the globe for a whole series of lavas extending back through the Quaternary and into the Pliocene (a period of some 10 Ma), the past Poles were all found to approximate to the present geographical Pole (Fig. 3.11). This supported the initial idea of the coincidence between geographical and magnetic Poles, and so it was reasonable to conclude that such coincidence had persisted throughout geological time.

A further complicating factor was the discovery that in some rock samples the magnetism was the complete opposite of that expected (180° reversal). In other words, instead of the remanent magnetism indicating that the magnetic North Pole was (at the time the rock formed) close to the

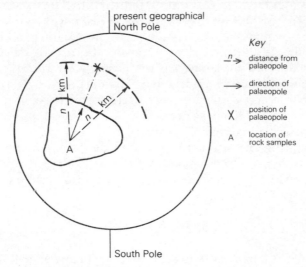

Figure 3.10 The location of a palaeopole.

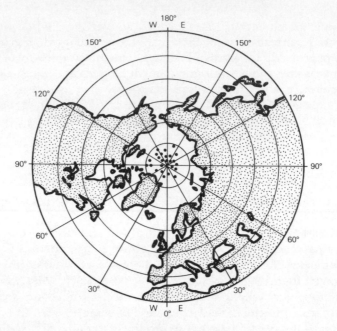

Figure 3.11 Palaeomagnetic poles of the Pleistocene and Pliocene.

geographical North Pole, as at present, it indicated that the magnetic North Pole was close to the geographical South Pole. Thus a north-seeking compass point would have pointed to the south at that time. This meant that either the magnetic field of the Earth had reversed, or those rocks contained some minerals that were capable of naturally adopting a remanent magnetism in the reverse direction to the Earth's field.

This problem was not resolved until the 1960s, but then it had the most important consequences, as will be seen in Chapter 6. Meanwhile, throughout the 1950s this did not prevent the use of declination and inclination values to locate palaepole positions. It could generally be decided on geographical grounds in which hemisphere a given area was at the time when some particular rocks were formed, so, if a palaeopole appeared reversed, all that was necessary was to reverse the set of readings 180° before determining the Pole position.

This situation in the late 1940s provided the basis for the subsequent explosive development of palaeomagnetic work, both on continents and in ocean basins, which will be discussed in the next three chapters.

SEISMOLOGY

In regions such as Japan, earthquakes are frequent and sometimes catastrophic events. An institute was established in that country in the 1880s to locate the sources of these earthquakes and if possible to predict their occurrence. From this institute the science of seismology developed. At that time it was realised that earthquakes were caused by some form of sudden disturbance within the Earth, so that, in addition to the obvious effects such as movements at the surface, vibrations must be propagated in all directions from the source, or **focus**. It followed that if a way of recording these seismic waves could be developed, earthquake foci could be located.

The necessary instrument, a **seismograph** (Fig. 3.12), was designed by a group of Englishmen (Milne, Grey and Ewing) who were foundation members of the Japanese institute. In essentials this instrument consisted of a heavily weighted boom pivoted against a massive support firmly attached to the ground. The heavy weight on the boom, because of its inertia, tended to remain stationary when everything else vibrated on the arrival of a seismic wave. The resulting relative movement between the boom and the rest of the instrument was recorded on a chart attached to a steadily rotating drum, by pens attached to the end of the boom. Such a recording is called a seismogram (Fig. 3.13).

The theory behind the propagation of seismic or shock waves pre-dated the invention of the seismograph by some 50 years, having been developed by Poisson in 1829. He showed that shock waves passing through a body are of two types: one causes particles to vibrate in the same direction as the

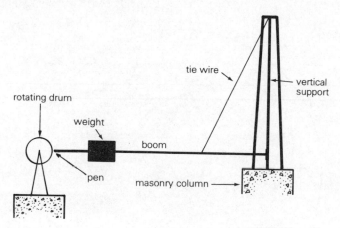

Figure 3.12 A Milne–Shaw seismograph.

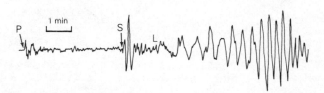

Figure 3.13 A seismogram.

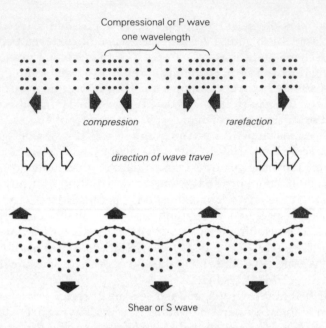

Figure 3.14 P and S waves.

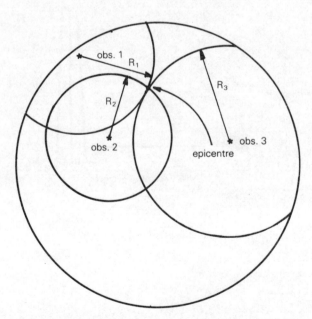

Figure 3.15 Location of epicentre deduced from arrival times of P and S waves at three observatories.

wave travels and are referred to as compressional, primary, or P waves; the other causes particles to vibrate at right angles to the direction of wave travel and are called transverse, secondary, shear, or S waves (Fig. 3.14).

P waves travel faster than S waves in all materials, but the actual velocities of both waves depend on the density and elastic properties of the materials involved. Both waves are refracted and/or reflected at boundaries between materials differing in these properties, as light waves are on passing from material of one refractive index to another. In addition, because the transmission of S waves is dependent on a shearing motion, they are not transmitted through liquids or gases, as these materials have no rigidity.

Oldham first identified Poisson's P and S waves in 1887 on the seismogram of an earthquake originating in Assam. It was then realised that, by knowing the arrival times of these two waves at three or more recording stations and assuming a relationship between the speed of transmission within the Earth and the distance travelled, it would be possible to determine the point on the Earth's surface directly above the focus of the earthquake, the **epicentre** (Fig. 3.15). Attempts were also made to estimate the depth of the focus of earthquakes, the principles behind which are outlined here (see box).

In the years subsequent to Oldham's breakthrough, there was an enormous amount of effort expended in establishing better values for both the travel times of seismic waves and the intensity of earthquakes. This resulted in the more exact determination of epicentres and focal depths. It was soon established that the overwhelming number of epicentres were located around the margin of the Pacific and on an east–west line through Indonesia, the Middle East and the Mediterranean (Fig. 3.16). Over 90% of them originated within 60 km of the surface, while the remainder occurred at various depths to a maximum of about 300 km.

Two major kinds of earthquakes were differentiated: tectonic earthquakes that result from the release of stress built up in rocks, and volcanic earthquakes resulting from gas explosions or the updoming and fissuring of volcanoes and their foundations. This last type almost always has a shallow focus and so the associated area of sensible disturbance is rarely more than a few hundred square kilometres. In contrast, tectonic earthquakes can have a much deeper focus, can release much greater energy and so can be detected over a much greater area of the Earth's surface.

With the recording of more and more earthquakes, the concentration of epicentres in the shaded areas of Figure 3.16 became even more marked. In addition, a rather broad band of low-intensity, shallow-focus earthquakes began to emerge along the Mid-Atlantic Ridge, which had been known as a topographic feature since the Challenger expedition of the 1870s (see Ch. 4). The most surprising development, however, had to do with the focal depth of earthquakes, for in 1922 some were reported as originating at a depth of 600–700 km. Initially such reports were rejected, ostensibly because of insufficient evidence, but the real reason was the general belief

Estimating depth of focus

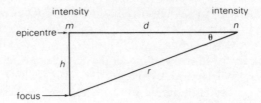

It was known that intensity decreases outwards from the focus inversely as the square of the distance, i.e.

$$intensity \propto 1/distance^2$$

So, knowing the intensity (m) at the epicentre and the intensity (n) at a known distance (d) from the epicentre, which is an unknown distance (h) above the focus, it follows that

$$n \propto 1/r^2$$

$$m \propto 1/h^2$$

and so

$$n/m = h^2/r^2$$

But $h/r = \sin \theta$, so

$$n/m = (\sin \theta)^2$$

from which θ can be found, because n and m are known. Also we have $h/d = \tan \theta$, so that

$$h = d \tan \theta$$

and as θ is now known, the depth of focus (h) can be determined.

that materials at this depth were not rigid enough to enable stress build-up and rupture to occur. As more and more evidence accumulated over the next 10 years, denial of the reality of deep-focus earthquakes became impossible, but, as they could not be ascribed to faulting, it was suggested that they could be generated by a sudden increase, or decrease, in the density of material at this depth, which would result in either a collapse or an explosion. However, more detailed analyses of seismograms by Japanese investigators in 1930 showed none of the expected differences between shallow- and deep-focus earthquakes, if the former were generated by

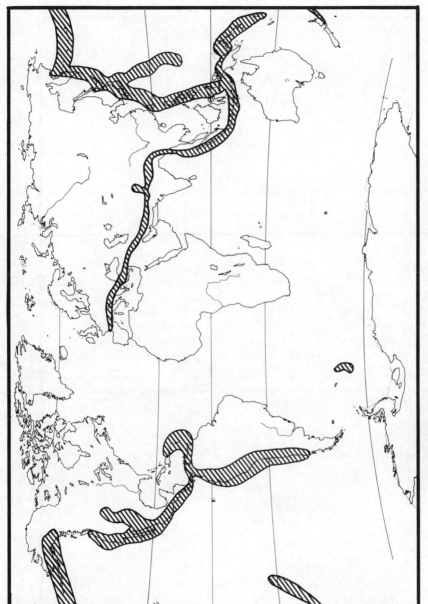

Figure 3.16 Major seismic zones (as recognised in the 1920s).

●●●●● deep-focus earthquakes ———— oceanic trenches

Figure 3.17. Deep-focus earthquakes and oceanic trenches.

faulting and the latter by collapse or explosion; in fact, it was highly probable that the causative mechanism was the same in both cases.

This inference was confirmed in 1936 when Beno Gutenberg examined all the evidence available and showed that deep-focus earthquakes were caused by differential movement along faults. In addition it could be demonstrated that, where sufficient data were available, these movements were in the same direction as movements causing associated shallow-focus earthquakes. It was also shown that practically all deep-focus earthquakes occurred on the landward side of oceanic trenches, particularly around the Pacific (Fig. 3.17). The great significance of this fact will become obvious in the next chapter.

In addition to locating places where the Earth was under stress, the study of seismic waves has also been largely responsible for our increasing knowledge of the internal structure of the Earth. Inferences could be made about the nature of materials encountered by seismic waves travelling within the Earth by combining the known travel times of P and S waves with the results of laboratory experiments on shock wave velocities in rocks of known densities and elastic properties, over a range of temperatures and pressures.

The process of interpretation is highly complex as there are so many variables, but it is still possible to appreciate something of the techniques employed. First of all a model of the Earth's interior is proposed; then the travel times of seismic waves through the model are calculated and compared with actual travel times: If they disagree the model is modified, the travel times recalculated and again compared with the actual times. The model is then adjusted again, if necessary, and the whole process repeated until the two sets of figures approach coincidence.

The application of this method can be seen in the way that a picture of deep Earth structures was built up in the first 40 years of this century. The simplest possible model of the Earth was one that had the same material throughout, with no variation in density or elastic properties, so that seismic waves would travel with constant velocities along straight-line paths (Fig. 3.18a). It was a simple matter to determine the expected arrival times of seismic waves at any point on the Earth's surface. When these results were compared with observed travel times, it was found that the calculated values were always greater than those observed and that this discrepancy increased the greater the distance from the epicentre. The seismic waves were therefore travelling faster than expected, a fact that could be explained by the waves taking a curved rather than a straight path through the Earth (Fig. 3.18b) and encountering faster travelling conditions at depth. Therefore the original concept of a uniform Earth had to be rejected. This was not surprising as it was already known, from gravity data and the Earth's moment of inertia, that the Earth's density increases markedly towards its centre. It was reasonable then to change the model to one in which the density increased with depth.

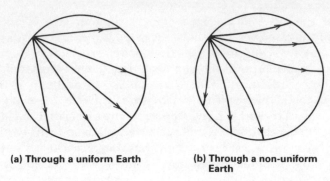

(a) Through a uniform Earth

(b) Through a non-uniform Earth

Figure 3.18 Seismic wave paths.

When this idea was examined, however, in the light of the velocity equations for P and S waves (see box), there were clearly some problems. Wave velocities are *inversely* related to the square root of the density and, all else being equal, increasing density at depth will cause a decrease in velocity and so longer travel times than are predicted. In fact, exactly the opposite occurred. The equations show, however, that there is a *direct* relationship between the velocity and the square root of the rigidity and also, in the case of P waves, with the incompressibility. The paradox could be resolved simply if these elastic properties have a greater effect on the velocity of seismic waves than does the density. Thus the Earth model that

---

Velocity equations of P and S waves

Velocity of P waves is

$$v_\mathrm{P} = \sqrt{\left(\frac{K + \frac{4}{3}\mu}{p}\right)}$$

Velocity of S waves is

$$v_\mathrm{S} = \sqrt{\left(\frac{\mu}{p}\right)}$$

Here p is density and K and μ are elastic properties: K is incompressibility (i.e. resistance to compression) and μ is rigidity (i.e. resistance to deformation).

It can be seen from the equation for v_S that if the rigidity μ is zero (that is, if the transmitting medium is a liquid or gas), the velocity of S waves will also be zero: in other words, S waves will not be transmitted.

[Prepared by F. Williams]

---

was adopted was one in which density increased with depth, but also one in which rigidity and possibly also incompressibility increased even more rapidly.

This model was found to agree with the travel times of the first arriving P and S waves up to an epicentral angular distance of 103°. From there to 143° none of the expected P or S waves were detected. Although seismic waves did occur beyond 143°, they were only P waves, not S waves. This pattern indicated a very sudden change in the character of the material traversed by seismic waves below a given depth within the Earth, which was very different from the gradual changes with depth noted up to this point. In 1906 Oldham interpreted this distribution of seismic waves in terms of a large, very dense spherical core. This would cause a marked decrease in the velocity of P waves so that they were refracted downwards and did not reach the surface until beyond 143°, leaving the intermediate **shadow zone** without direct waves (Fig. 3.19).

The complete lack of S waves beyond 103° meant that S waves could not be transmitted through the core material. The core then must have no rigidity at all; it must be liquid. The boundary appeared to be abrupt and its depth was determined as being at 2900 km by Gutenberg in 1914.

In 1936 the Earth model was further amended when Lehmann showed that the shadow zone was not completely free of waves, but that P waves of

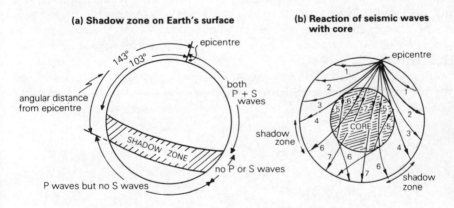

Figure 3.19 The shadow zone and how it is produced (see box).

Rays 1, 2 and 3 pass normally through the Earth along slightly curved paths. Ray 4 just glances the edge of the core without penetrating; ray 5, however, just penetrates the core and is sharply refracted downwards. On emerging from the core it is refracted again and reaches the surface almost on the oppo-site side of the Earth from the focus. Rays 6 and 7 are refracted by the core as shown; it can be seen that no ray can re-emerge within the shadow zone.

very reduced intensity were received within it. This could best be explained if there was a solid (i.e. rigid) inner region within the liquid core. This would refract the P waves upwards, and direct them into the shadow zone (Fig. 3.20).

By 1940 very accurate time travel tables had been compiled for P and S waves, mainly as a result of the work of Bullen and Jeffreys using the large volume of data steadily being collected from seismographs around the world. It was then possible to produce a graph showing the velocity of P and S waves as a function of depth (Fig. 3.21). By combining this with all other data available, variations in density, rigidity and incompressibility with depth (Fig. 3.22) could be inferred and used to produce a model of the deep structure of the Earth which is still generally acceptable.

As well as providing information on deep Earth structures, seismic observations also helped to elucidate structures nearer the surface of the Earth. This work was initiated in 1909 by Mohorovičić, who, when studying the records of a Croatian earthquake, found that two distinct sets of P and S waves were received. This was interpreted as being due to one set of waves travelling directly to the receiving station through an upper layer of the Earth, while a second set had travelled through a lower layer at a greater speed to reach the station first (Fig. 3.23). It was soon realised that this doubling of the P and S waves was a frequent occurrence on seismograms from stations within 800 km of the epicentre of any earthquake and indicated that the boundary between the two layers of material responsible for the velocity difference was within about 50 km of the surface. Also the generality of its occurrence showed that it had a world-wide distribution. This seemed to be a most elegant confirmation of the concept of isostasy as developed by Airy and Pratt, and so in conformity the upper layer was referred to as the crust and the lower layer as the mantle. The actual travel times of the seismic waves through the two layers indicated that the crust

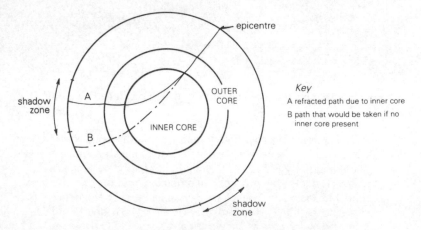

Figure 3.20 The detection of the inner core.

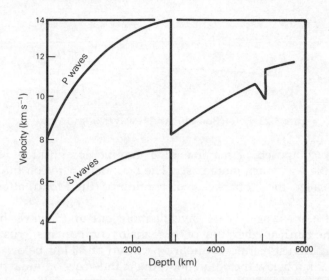

Figure 3.21 Variations in velocity of P and S waves with depth.

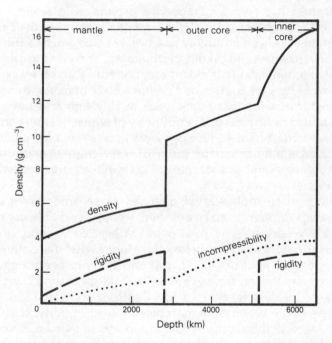

Figure 3.22 Variations in physical properties with depth.

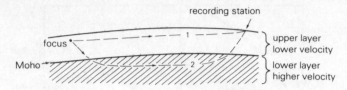

Figure 3.23 Refraction of P and S waves at the Moho.

was mainly composed of material of a granitic composition while the denser mantle was much more basic. The boundary between the two has since been called the Mohorovičić discontinuity (the Moho) after its discoverer.

Thus, in the space of 40 years, investigations of P and S waves built up a picture of the Earth as consisting of three major components: crust, mantle and core. The crust/mantle boundary occurred at 35 km below the continents and 6 km below the ocean floor; the mantle/core boundary was at a depth of 2900 km and the liquid/solid core boundary was 5100 km below the surface.

Since 1940 there has been a tremendous expansion in seismology and most of this has been concerned with the near-surface zone of the Earth. For the most part this has been for highly practical reasons, such as the need to detect nuclear explosions and predict earthquakes, as well as to assist in the search for oil and minerals. This meant that not only P and S waves (called **body waves** since they are transmitted within a body of material) needed to be investigated, but also other types, such as Rayleigh and Love waves, which are confined to surfaces of discontinuity of which the most important is at the Earth's surface. These developments necessitated the recording of seismic waves over a much greater range of wavelengths, so that a whole battery of seismographs was developed as well as great networks of receiving stations.

The resulting more sophisticated analyses of seismic waves enabled much more subtle differences to be detected. Thus a layer of lower rigidity, the so-called low-velocity zone, was detected between depths of 60 and 250 km below the surface, that is below the Mohorovičić discontinuity and within the upper part of the mantle. Its presence had first been suggested as early as 1926 by Gutenberg, from evidence that seismic waves travelled 6% slower in a layer of material centred at a depth of about 150 km. This he thought indicated the presence of materials of lower rigidity, that is more capable of flowage, at this depth. This opinion was not well received at the time and was generally dismissed on grounds of insufficient evidence. It was not until 1960 that convincing evidence was finally produced. On 22 May of that year one of the most violent earthquakes of the century affected Chile and this was so strong that the whole of the Earth vibrated as a unit; it rang like a bell. For the first time seismographs that could record the very long-period earthquake waves from such a shock were available and their

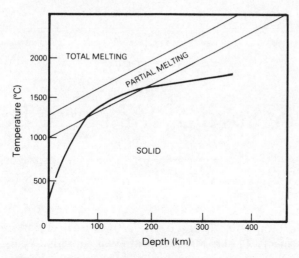

Figure 3.24 Possible explanation of the asthenosphere.

records were enough to convince Earth scientists that this less rigid layer was a reality.

It was interpreted as being a zone between 80 and 200 km below the surface in which partial melting occurs as a result of the increase in temperature with depth outstripping the inhibiting effect of increased pressure (Fig. 3.24). Such a zone would be rather weak mechanically and so be able to flow. It was therefore called the **asthenosphere** to distinguish it from the overlying much more rigid **lithosphere**. This physical differentiation is very different in both character and location from the inferred compositional difference between the crust and mantle, referred to previously. The recognition of the asthenosphere was one of the fundamental contributions of seismologists to the new model of the Earth developed in the 1960s; others arose from investigations in the ocean basins and they will be considered in the next chapter.

CHAPTER FOUR

The geology of the ocean basins

Ocean basin geology is a creation of the 20th century and more particularly of the years since 1950. The only exception was provided by the exploration of oceanic trenches. These are narrow, elongate, sharply demarcated depressions, up to 10 000 m deep, which had been studied in some detail. By the 1920s it was known that they were distributed around the margins of the Pacific, south of Indonesia and north of the Caribbean arc of islands and were in such close association with the distribution of earthquake epicentres as to suggest a genetic connection (Fig. 3.17). The probing of this possibility resulted in the first real oceanic geological investigation when Vening Meinesz, a Dutch geophysicist, measured variations in the Earth's gravity field over the trenches.

As will be recalled from the discussion in Chapter 1, the most accurate determinations of gravity require the rate of swing of a pendulum to be timed with great precision. For such measurements, conditions of great stability for the pendulum mounting are necessary. Such stability was not achieved at sea until Vening Meinesz, in the 1920s, made a series of measurements in a submerged submarine while on a voyage from Holland to Java via the Panama Canal. For the first time it was directly demonstrated that gravity values at sea are much the same as on land, which meant that much of the ocean floor is in isostatic equilibrium, so that the difference in height between continents and ocean basins must be explicable in terms of much denser rocks underlying the ocean basins. All of this had already been worked out indirectly by Pratt (see Ch. 1) 40 years before. However, when the submarine was above an oceanic trench near to what is now Indonesia, the measured gravity was found to be markedly lower than expected, indicating an absence of isostatic equilibrium. The effect was as though a great mass of material was missing from beneath the gravity meter, or as though some less-dense material had replaced material of normal density at depth. Mapping the extent of this anomalously weak gravity zone established it to be 160 km wide and 8000 km long, following the pattern of trenches that are parallel to and south of the island arc. It curved north with the trend of the arc to pass west of New Guinea and up the east coast of the

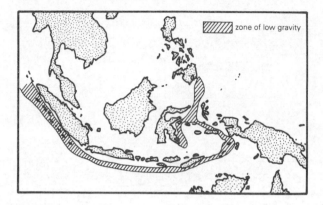

Figure 4.1 Negative gravity anomalies of Vening Meinesz.

Philippines (Fig. 4.1). Similar gravity features were also discovered associated with the trenches east of Japan and those adjacent to the Caribbean island arc.

As was seen previously, one great weakness of Wegener's hypothesis of continental movement was that the external forces that he suggested as possible mechanisms were extremely small. Arthur Holmes saw in the gravitational and seismic data from the trenches evidence of an internal Earth force of sufficient strength to move continents laterally. He suggested that these movements could be the result of mantle convection currents (Fig. 4.2) that rise and diverge under a continental mass. The resulting tension is eventually able to cause the continent to split and the fragments to move apart. In the space between the drifting continents a new ocean basin would form, floored by basaltic lavas erupted in the process. The basin would grow with time, on either side of the mid-oceanic ridge that marks the site of the original split. Where two of these currents meet and turn downwards into the mantle, a trench would be formed at the surface. This idea of Holmes also provided a solution to another of Wegener's problems, for on this basis the continents did not have to move through the denser substratum. They were carried piggy-back on the convecting basaltic layer, and fold mountains formed where they abutted against the trenches.

Among the few to capitalise on this idea was Vening Meinesz, who attempted to give the process a historical dimension. He suggested that when the Earth was young there was only one convection cell (Fig. 4.3a). Over the area of the descending current there would be a surface accumulation of the less-dense material to form a single continental mass (Pangaea). With time, the core gradually developed to such a size that the simple circulation pattern was no longer possible and it was replaced by a number of smaller cells (Fig. 4.3b). In this new situation some of the rising currents came up under the continental mass and tore it apart, forming a new ocean on either side of a central ridge. As the core continued to grow, the

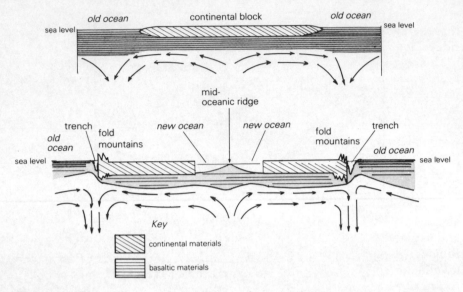

Figure 4.2 Convection current mechanism for continental movement, as suggested by A. Holmes.

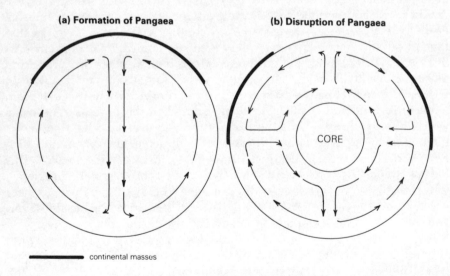

Figure 4.3 Convection currents and continental formation according to Vening Meinesz.

convection pattern was further constrained and more new cells evolved, so that the patterns of continental movement changed and new mid-oceanic ridges developed.

This model of macroscale Earth processes was the first to suggest how both tensional and compressional phenomena could occur at the same time on the Earth's surface. Furthermore, by its emphasis on the role of the mid-oceanic ridge, it was the first sign of geologists moving away from a continental towards an oceanic viewpoint. However, at the time these ideas were dismissed as being highly speculative, with little or no basis in fact. In view of the remarkable similarity between them and the ideas adopted 40 years later, they have retrospectively taken on a considerably enhanced significance.

At this time the ocean basins were thought to consist of a featureless abyssal plain at a depth of about 5 km, interrupted by occasional trenches and in places by broad rises, of which the best known was the Mid-Atlantic Ridge. This view persisted until the mid-1940s when the echo sounder, invented in the 1920s and perfected in World War II, enabled a continuous record of ocean-floor topography to be obtained.

The first results of these records, published in 1946 by Professor Hess of Princeton, showed that the abyssal plain had considerable topographic variability. The data were obtained by Hess when Commander of a US Navy transport ship during the War. The echo sounder, normally only used in shallow water, was kept operating while the ship was crossing the central Pacific between the Marianas and Hawaii. Twenty flat-topped peaks, or **guyots**, were located rising 3–4 km from the abyssal plain (to within 1–2 km of the surface) and the presence of another 140 were inferred from marine charts of the area.

In trying to explain these features, Hess was constrained both by the concept of the permanence of continents and ocean basins (particularly the great age of the Pacific) and by the notion that movement could only take place in a vertical sense. In addition, it was necessary to explain why, if the flat tops of the guyots were the result of marine planation at sea level, corals had not established themselves on these submerging platforms to develop atolls, as originally suggested by Charles Darwin (see box). Hess offered the extremely tentative explanation that planation had occurred in Precambrian times when there were no corals and subsequent guyot submergence was the result of the deposition of sediment that had progressively raised sea level throughout the Pacific, so that those guyots furthest from the surface were the oldest.

Subsequent work in the same area by the Scripps Oceanographic Institute of California showed that the guyots were the peaks of submarine mountain ranges formed by the extrusion of basic lavas and pyroclastics through tensional fissures and faults in the crest of a low swell. In fact the whole area could be seen as the submerged equivalent of the Hawaiian island chain. As shallow-water Cretaceous fossils were recovered from some guyot

Darwin's theory of atoll formation

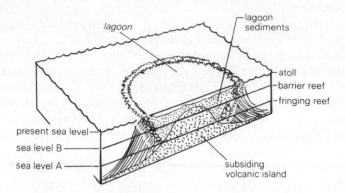

Darwin suggested that, as a volcanic island slowly subsided, the surrounding coral would keep pace with the subsidence, growing upwards and outwards.

Stage 1: Volcanic island, sea level A. The island is surrounded by a fringing reef, where the coral grows close to the shore.

Stage 2: After some subsidence to sea level B, the coral has grown up and outwards to form a barrier reef, away from the island, and the slope between the island and the reef is filled with sediment.

Stage 3: The island has now sunk below sea level, the coral forms a complete ring and there is no central island visible. An atoll has been formed.

summits, the submergence must have occurred since that time. Again working within the constraint that only vertical movement was possible, it was postulated that guyot submergence resulted from the isostatic adjustment of the oceanic crust to the additional loading imposed by the volcanic materials. This could have been quite rapid once the bearing strength of the crust was exceeded, so that coral growth could not keep pace and hence atolls were not formed. As will be seen in Chapter 7, plate tectonics finally vindicated Darwin by providing a rational explanation of the great amount of submergence required.

The Scripps Institute was also responsible for a much larger survey of the northern Pacific abyssal plain, covering some 20×10^6 km^2, the results of which were published in 1955. This established that most of this enormous area had been affected by fracturing and volcanism. In particular, four major fracture zones were mapped, having widths up to 100 km and following parallel routes for between 2400 and 5600 km. The most outstanding feature of these fracture zones was the presence of tremendous escarpments, up to 1.7 km high and 1700 km long (Fig. 4.4). These features were tentatively explained in terms of the drag effect of mantle convection currents, which caused the thin and brittle oceanic crust to be cracked and distorted without being moved laterally to any great extent. This extremely tentative solution was short-lived in the face of an investigation into variations in the magnetic intensity possessed by the ocean-floor rocks in the eastern part of the area crossed by these fracture zones.

The total intensity of the Earth's magnetic field was known to vary from 0.3 Gs (gauss) near the Equator to 0.7 Gs in polar regions. The variations in magnetic intensity at the sea surface resulting from the magnetism of the ocean-floor rocks are anything from 1000 to 100 000 times smaller than these already small values. To measure such low intensities in the presence of the Earth's field presented severe technical difficulties, for the degree of instrumental sensitivity required was very high indeed. Development of such an instrument began in World War II when it was realised that submarines could be detected magnetically. The proton-precession magnetometer was developed with a precision of 1 in 50 000 and used as a magnetic airborne detector. It was lowered from a plane to trail below and behind, well clear of any magnetic disturbance produced by the plane, and it did effectively detect submarines at shallow depth. After the War it was realised that this type of magnetometer, suitably waterproofed and towed far enough behind a ship to be free of the ship's magnetic disturbance, might provide information about the magnetic properties of the ocean floor. This

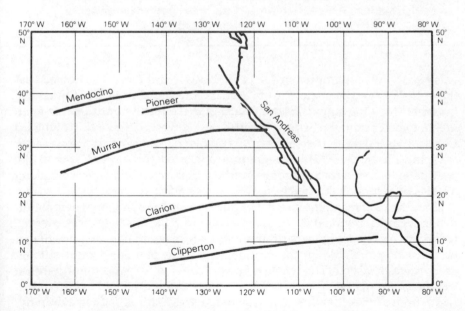

Figure 4.4 Fracture zones of the north-east Pacific.

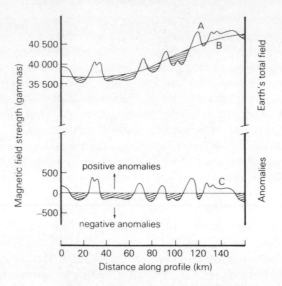

Figure 4.5 Magnetic anomalies.

Curve A is the observed total field strength along a certain profile, expressed in gammas, that is 1/100 000 of a gauss.

Curve B is the average total field strength along the same profile, i.e. regional magnetic intensity.

Curve C shows the anomalies resulting when B is subtracted from A, which are either positive or negative depending on whether A is greater or smaller than B.

was successfully attempted in the early 1950s and the results showed that the rocks of the ocean floor were remarkably variable in their magnetic intensity. Nothing significant was seen in these results until 1955 when a magnetometer from the Scripps Institute was towed behind a ship of the US Coast and Geodetic Survey, which was involved in an intensive deep-water mapping programme. This programme included the eastern part of the previously mapped fracture zones, and magnetic data were collected along a series of accurately located east–west tracks 8 km apart.

From the total magnetic intensity figures recorded by the magnetometer, the appropriate regional magnetic intensity (the Earth's field) was subtracted to leave a residual **magnetic anomaly**, which could be either positive or negative (Fig. 4.5). When these anomalies were plotted, a most striking pattern was revealed. The whole map was covered by parallel north–south trending, alternating positive and negative anomalies, which varied in width from a few kilometres to tens of kilometres. Although their continuity was interrupted by the east–west trending fracture zones (Fig. 4.6), it was possible to match the anomalies on either side by invoking a considerable degree of lateral shift. Thus a match of the anomaly pattern could be achieved by moving the north side of the Pioneer Fault Zone 265 km to the east, and in the case of the Murray Fault Zone by moving the south side 154 km to the east. No such match was possible across the Mendocino Fault Zone until the survey area was extended another 20° of longitude to the west, when a match was achieved by moving the north side of the fault 1160 km to the east (Fig. 4.7).

This offsetting of the magnetic patterns was interpreted as being the result of transcurrent or wrench faulting, although in these cases the amount of suggested movement was very much greater than for faults of this nature located on the continents. Furthermore, seismic evidence suggested that all such movement must have taken place at some time in the past, for the newly acquired ability to locate more exactly the epicentres of

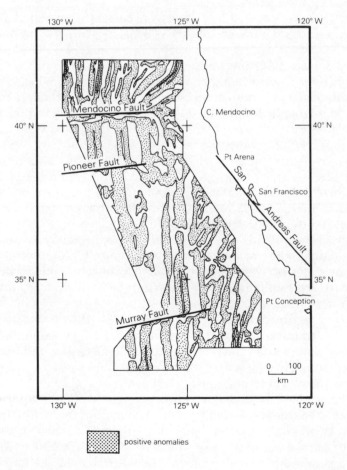

positive anomalies

Figure 4.6 Magnetic anomaly patterns of the north-east Pacific.

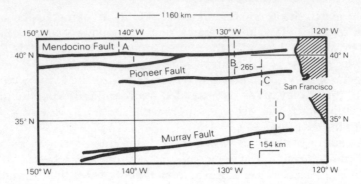

Figure 4.7 Matching of magnetic anomaly patterns (A, B, C, D and E mark positions of matching patterns).

oceanic earthquakes showed that none were currently generated along the fracture zones, despite the occurrence of what appeared to be huge fault scarps.

Thus by the late 1950s investigation of the Pacific Basin had established the presence of a great number of features, none of which could be satisfactorily explained. Other problems arose when, at about the same time, the Mid-Atlantic Ridge was investigated by members of the Lamont Geological Observatory led by Maurice Ewing. Although in 1953 there were only six topographic profiles across the North Atlantic, by 1959 several dozen were available, and from these it was apparent that Atlantic topographic diversity was every bit as great as in the Pacific. From the profiles the topography of the Mid-Atlantic Ridge could be mapped precisely, and this showed that the ridge was bisected axially by a massive rift valley up to 2 km deep and from 14 to 60 km across.

The significance of this rift valley was not immediately apparent, but when the much more precise earthquake epicentre locations were plotted on the same map they were found to be almost entirely confined within the zone demarcated by the rift valley instead of being distributed across the ridge, as the old data indicated (see Ch. 3).

This was a fact of tremendous significance, for the coincidence of rift valleys and earthquakes pointed to tectonic activity. The Lamont scientists re-examined and re-interpreted all the old earthquake data and in addition obtained new data from seismograph stations that they established in South Africa and India where previously data had been sparse. Their results showed the narrow earthquake zone to be world-wide (Fig. 4.8). It ran from north to south for the whole length of the Atlantic and then turned east and north-east between Africa and Antarctica, where it divided. One branch went north and north-west through the Indian Ocean and Gulf of Aden into the Red Sea and African rift valleys. The other continued its eastward course and passed between Australia and Antarctica before swinging

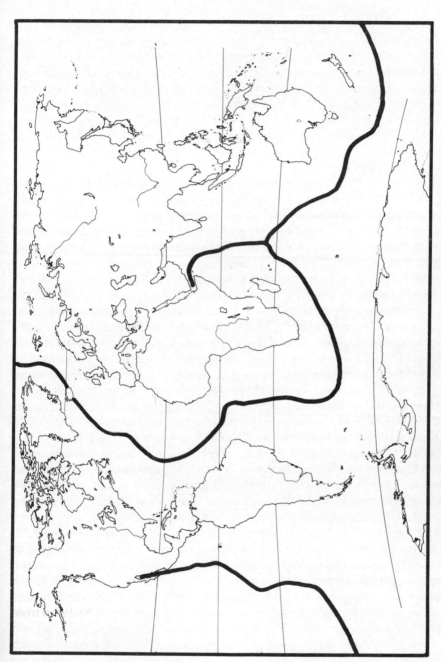

Figure 4.8 The extent of the mid-oceanic ridges as inferred from earthquake epicentres.

north-east across the Pacific to come ashore in California along the San Andreas Fault.

Despite the fact that coincidence between the rift valley and the narrow seismic zone had only been established over a few hundred kilometres in the North Atlantic, Bruce Heezen, one of the Lamont scientists, took a great leap into the unknown and stated that wherever this narrow zone of earthquake epicentres occurred in the oceans of the world it would be coincident with the rifted crest of a mid-oceanic ridge. This was confirmed when a Scripps expedition to the South Pacific found a ridge exactly where Heezen had predicted, in a part of the ocean where the narrow line of earthquakes was mapped, but where no ridge had previously been observed or even suspected.

Scientists at Lamont tested Heezen's idea in the Indian Ocean and, using the earthquake epicentres as a guide, made eight crossings of a previously uncharted ridge and rift, every bit as topographically spectacular as the Mid-Atlantic Ridge. Every time the idea was tested it was found to be correct and, in the face of such unanimous confirmation, it was quickly accepted as an established fact by most geologists. However, there was no unanimity as to the cause of such a narrow but world-encircling zone of tension in the Earth's crust.

Still further problems were raised by the Lamont group's investigation into the nature of the oceanic crust. This involved the detection of artificially generated seismic waves after they had travelled through and been refracted by the different layers below the ocean floor; this was a method pioneered in the 1930s by Maurice Ewing on the continental shelf and slope off the south-east coast of the USA. As with echo sounding, the early methods were extremely cumbersome and somewhat hazardous, for depth charges and grenades were used to generate the seismic waves. After 1945 this method was gradually replaced, except where very deep penetration into the crust was required, by the development of non-explosive noise makers, such as sparkers and airguns. These produce low-frequency sound waves that are able to penetrate deep into the ocean floor, and such devices made investigation of the layers beneath the ocean floor hardly more difficult than using an echo sounder. An overall picture of the oceanic crust was emerging by the 1950s, which in several ways was quite unexpected.

Compared to the structural and compositional heterogeneity of the continental crust, the oceanic crust appeared to be amazingly uniform, consisting almost everywhere of only three layers. The top layer (layer 1) had an average thickness of 0.4 km, although this varied considerably and in some areas was even completely absent. Its slow rate of shock wave transmission indicated a sedimentary composition. Beneath layer 1, layer 2 was 1–1.5 km thick and its rate of wave propagation indicated a basaltic composition. Layer 3 was 5 km thick with seismic indications that it was made up of denser, probably ultrabasic, rocks. The upper surface of layer 2 was found to be extremely uneven, whereas the overlying sediments of layer 1 showed no evidence of deformation. The persistence of ideas about the permanence of continents and ocean basins caused most geologists to consider the rugged surface of layer 2 to be the primordial crust of the Earth and the overlying undisturbed sediments of layer 1 to be the result of continuous gentle sedimentation since that time. It was thought that deep drilling in the ocean basins would be very informative because it would yield a complete history of oceanic sedimentation from layer 1 and a knowledge of the primordial crust from layer 2. Also the Mohorovičić discontinuity lay at a depth of only 7 km beneath the ocean floors, compared with about 35 km below the continents, which made the direct sampling of mantle material a distinct possibility. Such deep ocean drilling was attempted in the ill-fated Mohole project, which was initiated in 1959, but abandoned in 1966 amid a welter of political and cost problems. As will be seen later, however, the work accomplished in this period had important implications for the subsequent exploration of the ocean basins.

Some other discoveries about the nature of the deep-ocean-floor sediments seemed to be completely inexplicable as long as the primordial nature of layer 2 was maintained. It was found that the average thickness of layer 1 in the Pacific was 300 m and in the Atlantic 600 m, yet, if this layer was the result of sedimentation throughout geological times, it should have been from 3000 to 6000 m thick. Furthermore, it was to be expected that such a layer would be uniformly distributed across the deep-ocean floor, whereas it was generally absent near the mid-oceanic ridges and gradually increased in thickness away from them.

As this discussion indicates, during the 1950s ocean basin geology had produced so many unexplained observations that by 1960 geologists were feeling rather uneasy about the fundamentals of their subject. This state of general uneasiness was increased by concomitant developments in continental geology, and these will be the subject of the next chapter.

CHAPTER FIVE

The impact of palaeomagnetism

It was during the 1950s that fundamental ideas concerning continental geology began to be questioned. Much of this questioning came from the greatly increased number of geophysicists, particularly in Britain, engaged in investigations of palaeomagnetism. This great increase in palaeomagnetic research was mainly due to one man, Professor P. M. S. Blackett (a Nobel Laureate in physics and later Lord Blackett, President of the Royal Society). Blackett became involved in palaeomagnetism in the late 1940s as a result of his more fundamental work on the problem of how the Earth generates and maintains its magnetic field. It seemed that the past history of the magnetic field, as provided by palaeomagnetic investigations, could help in elucidating this problem.

The intervention by a man of such scientific eminence resulted in greater research effort being directed into this previously fairly neglected area, so that within a relatively short time the difficulties hindering the full development of palaeomagnetism had been resolved. A magnetometer of extreme sensitivity was designed that could accurately determine the very weak fields involved in rock magnetism. In addition, statistical techniques were developed to achieve reliable mean determinations of both declination and inclination from a number of samples.

This work contributed little to solving the original problem of the generation of the Earth's magnetic field, but by the time this was realised palaeomagnetic studies had taken on a life of their own. It became possible to locate the position of rocks in relation to the Pole at particular times in the past and this could be used as a quantitative test of Wegener's ideas of continental movement.

In 1954 an investigation was concluded on the chemical remanent magnetism of 500 samples of Triassic sandstone from nine sites in England. For five sites (Fig. 5.1) the mean declination was N29°E and the mean inclination 34° downwards, while for the other four sites the mean declination was S39°W and the inclination 16° upwards. As the diagram shows, these two sets of figures are almost 180° removed from one another, being an example of the magnetic reversals referred to in Chapter 3. Even though

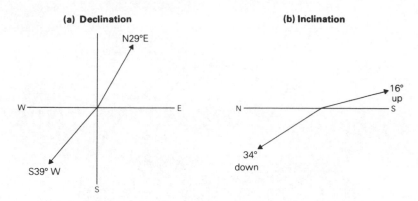

Figure 5.1 Palaeomagnetism of Triassic sandstones.

the cause of such reversals was unknown at that time, there was sufficient evidence to locate the British Isles in the Northern Hemisphere in the Triassic. The first set of figures (N29°E, 34° down) could therefore be regarded as reflecting normal magnetisation and the second set (S39°W, 16° up) reversed magnetisation. The Pole, in relation to N29°E, 34° down, was then at 50°N, 146°W (in the North Pacific), very far removed from its location at present and for the last 10 Ma. This was of course the geographical Pole, because it had already been concluded that the magnetic and geographical Poles had statistically coincided throughout geological time (see Chapter 3).

Within a year of these initial findings, enough results were available from the rocks of the British Isles and western Europe to determine the Pole position from the Precambrian to the present (Fig. 5.2). This showed that, in terms of present-day geography, the Pole was located near the present west coast of North America in the later Precambrian. It had moved to the middle of the Pacific by the Cambrian, passed by the northern part of Japan in the Carboniferous and then proceeded across Siberia to its present position. Such movement meant that Britain and western Europe must have originally been somewhere south of the Equator and passed through the equatorial zone and the subtropics to reach their present temperate situation. A certain part of this pattern appeared to substantiate the kind of palaeoclimatic evidence used by Wegener. The presence of thick coals and coral-bearing limestones in the Carboniferous rocks of Europe and of evaporites and desert sandstones in the Permian rocks are consistent with the palaeomagnetist's estimate of latitudes in those periods of between about 20°N and 20°S. The presence of large reptiles and reef-building corals in the Jurassic is also consistent with an estimated latitude 10–20° nearer the Equator than today.

It was possible to explain such events either by movement of the Poles themselves, or by the independent movement of the continents. Wegener had invoked both types of movement and, as was seen in Chapter 2, both

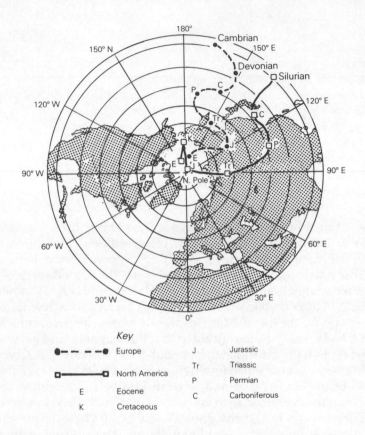

Figure 5.2 Polar wander paths for Europe and North America.

were rejected as being geophysically impossible. The geophysicists now found themselves in the same situation as Wegener; they were faced with undoubted evidence of movement but had no way of explaining it. They took the least unpalatable possibility and plumped for polar wandering, for the idea of the permanence of continents and ocean basins was still very strong. Their decision was made somewhat easier because just at this time (1955) some astronomers reported evidence indicating that the Earth was not as rigid as had previously been thought. In these circumstances they considered polar wandering could occur, although the suggested mechanisms remained highly speculative.

The acceptance of polar wandering as an explanation of the palaeomagnetic results had an immediately testable consequence. If only the Poles had moved, every land mass of the world would have the same polar wander path, whereas if the continents had moved relative to one another they would all have different polar wander paths. The first palaeomagnetic results from the Triassic of the southwestern USA seemed to support the idea that the Poles had moved, for at first a common Pole with the English Triassic rocks was indicated. However, as determinative methods became more exact, it became apparent that this was not so, as the North American Triassic Pole was located at least 20° to the west of the British Pole.

The situation became much clearer after samples had been collected from the huge sequence of rocks exposed in the Grand Canyon. These allowed the Pole position of North America to be determined from Precambrian times onwards, as had been done for Britain and western Europe. This showed that all American Poles lay 30° west of those from Europe, between the Precambrian and the Triassic (Fig. 5.2), giving for this period two distinct but parallel polar wander curves. It was found that, by shifting the American continent eastwards by 30° for this period, the two polar wander paths could be made to coincide. This would remove the Atlantic Ocean and America would lie alongside Europe. These two polar wander paths could be interpreted as indicating that from the Precambrian to the Triassic the two continents belonged to a single land mass and that during the Triassic they separated and moved to their present positions. In other words, these results were direct proof of continental movement.

Convincing as this appeared to be, there was a very large hole in the argument. As was well known, palaeomagnetic determinations only give a latitude and a Pole direction and say nothing about the degree of longitudinal separation between two points. These two polar wander paths could therefore be explained by the two continents maintaining their present degree of separation throughout geological time, but rotating relative to one another. Moving North America 30°W since the Triassic was only one possible solution out of many and it had been selected only because it confirmed what Wegener had said.

Better evidence for relative continental movements, however, was provided by the palaeomagnetic results from the southern continents, particularly those from the basaltic lavas of the Indian Deccan. These gave a Jurassic Pole position of 20°N, 69°W and a lower Tertiary Pole at 53°N, 83°W, compared to corresponding figures for Europe of 58°N, 122°E and 76°N, 150°E. Such figures can only be reconciled by differential movement between these continents during this period.

Inclination figures alone from these same Deccan basalts give a very good idea of the rate of movement of peninsular India. They varied from 64° upwards in the Jurassic to 26° upwards in the lower Tertiary and 17° downwards in the middle Tertiary, which indicated that India must have moved from the Southern to the Northern Hemisphere over a distance of some 7000 km in 70 Ma. Such a high rate of movement, over such a great distance, is exactly what had been demanded by Wegener to explain the fact that India had been glaciated in the Permian and was involved in creating the Himalayas in the late Tertiary. Further, as more results became available not only from India but also from Australia, Africa and South America, it was found that they all had very different polar wander paths, but that many of the differences could be reconciled when these continents were reassembled into Gondwanaland (Fig. 2.3) along the lines suggested by Wegener and du Toit; all of which suggested that Wegener had been right after all!

However, even if this was accepted, the fundamental problem remained of how it had happened and how the continents had moved through the ocean basins, a movement that now seemed all the more impossible because the presence of the immense mid-oceanic ridge indicated an oceanic crust of considerable strength. A way out of this dilemma was suggested by Professor S. W. Carey, of Hobart, who, in 1954, postulated that the continental masses had completely covered the surface of a much smaller Earth in Palaeozoic times. The Earth had since expanded, fragmenting the continental cover and creating the ocean basins in between from mantle material that reached the surface through expansion cracks now marked by the mid-oceanic ridge. By this means, although the distance between the continents has increased, they have not moved in relation to the mantle rock beneath them. Carey, a structural geologist and long a believer in Wegener's ideas, became convinced as a result of a great deal of detailed work on continental reconstructions that this was the solution.

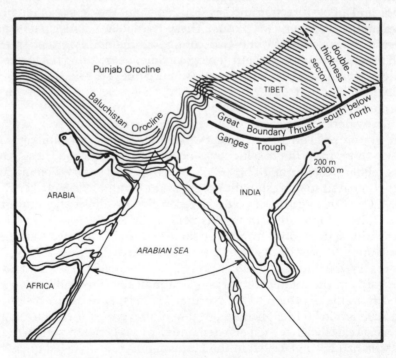

Figure 5.3 The Baluchistan Orocline and the Arabian Sea (after Carey).

Carey recognised that certain major features of the Earth's surface were produced by continental movement. In places the fold mountain belts of the world appeared to have been bent very sharply, for example the mountain ranges of southern Iran and Pakistan (Fig. 5.3).

This Baluchistan flexure or **orocline** was postulated as having been formed by the opening of the Arabian Sea during the separation of India and Africa. By reversing such processes and using the 200 fathom (~366 m) contour as the line of junction, Carey was able to achieve very precise fits between continents. Despite this success, he could not reassemble Wegener's Pangaea without leaving very large gaps between areas that geological evidence suggested should have been closely adjacent. However, if the Palaeozoic Earth had had a much smaller radius, all these problems disappeared and subsequent Earth expansion would explain the present disposition of the continents without having to appeal to continental movement across the mantle.

While this novel suggestion caused considerable discussion, there were several reasons why it never gained more than a few adherents. Calculations by astronomers of the amount of expansion that could have occurred, if there had been a gradual change in the gravitational constant since the origin of the Earth, showed this to be far too small to account for the amount of continental dispersion during the Mesozoic demanded by the palaeomagnetic results.

This conclusion was further supported by other palaeomagnetic results. For example, if a continent had remained a unified mass through such a process of expansion, it would subtend a smaller angle at the centre of the Earth after this process had occurred than before (Fig. 5.4); in other words,

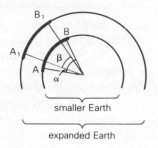

Figure 5.4 Palaeomagnetic test of the expanding Earth.

On a rigid continental block of a smaller Earth, two locations A and B would subtend an angle α at the centre of the Earth. On expansion, the rigid continental block would remain the same size, so that the same two locations (now A_1 and B_1) would also remain the same distance apart but the angle subtended at the centre of the Earth by them (β) is smaller than α.

the latitudinal distance between places on the continent would have decreased. As palaeo-latitudes can be derived from palaeo-inclination values, this could be tested by determining the inclinations of rocks formed before the postulated expansion from sites situated as far apart on the continental mass as possible and comparing the results with the present differences in latitudes. There was found to be little or no difference between past and present values, and, while the accuracy of the results was not enough to disprove expansion completely, they were good enough to rule out anything like the rate and amount of expansion required by Carey.

The most important reason for the demise of the expansion theory, however, was the revival of Holmes' idea of thermal convection currents in the mantle (Fig. 4.2). A relatively small, but highly influential, group of geologists and geophysicists decided that there could no longer be any doubt that the results from palaeomagnetism indicated that continental movement had occurred and that, of all the proposed causes, mantle convection provided the only really plausible mechanism. This revival was initiated by R. S. Dietz (1961) and H. Hess (1962) who saw in the piggy-back transport of continents a sufficient explanation of the continental palaeo-magnetic results. The concomitant formation of ocean basins in the wake of this continental movement meant that, rather than being primaeval Earth structures, ocean basins were of recent formation and relatively ephemeral features compared to the continents. The derivation of the oceanic litho-sphere from a single source, the mid-oceanic ridge, explained the uniform-ity of layers 2 and 3, while the youthfulness of the ocean basins accounted for the general paucity of layer 1 sediments; the increase in thickness of this layer away from the mid-oceanic ridge could be seen as a consequence of the increasing age of the ocean floor in the same direction. The highly appro-priate term **sea-floor spreading** was coined for the whole process.

The rather bare bones of this model were filled out considerably by two leading geophysicists, Sir Edward Bullard of Cambridge and J. Tuzo Wilson of Toronto. Bullard had been struck by the exactness of Carey's continental reconstructions but was aware of the accusations of subjectivity which had long been used against this kind of work. He removed much of the basis for these criticisms by making use of a computer to fit together the continents on either side of the Atlantic, by providing it with data on continental shape at 50 km intervals at depths of 100, 500, 1000 and 2000 fathoms (~183, 915, 1830 and 3660 m). By this means a best fit was established as occurring along the 500 fathom (~915 m) contour, which coincides with the maximum slope of the continental rise (Fig. 5.5).

The excellent results achieved by the application of this objective method established it as a reliable means of continental reassembly. This made it possible to look for geological continuity between coastlines that had apparently once been joined, in a much more detailed and quantitative manner than had been possible for either Wegener or du Toit. Typical of

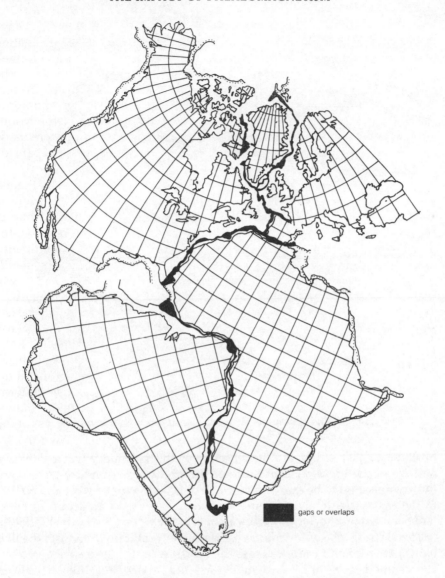

gaps or overlaps

Figure 5.5 Computer fit of Atlantic continents (after Bullard).

such work was the use of radiometric dating techniques to achieve correlations between West Africa and South America (Fig. 5.6). In West Africa a
sharp boundary had been established between a 2000 Ma geological
province in Ghana, the Ivory Coast and countries to the west and a 600 Ma
province in Dahomey, Nigeria and to the east. This boundary heads in a
southwesterly direction into the Atlantic near Accra in Ghana. If Brazil had
been joined to Africa 600 Ma ago, the boundary between the two provinces
would occur in South America close to the town of Sao Luis on the

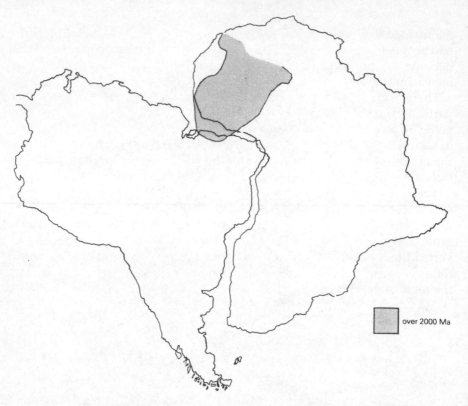

Figure 5.6 Continental fit confirmed by radiometric dating.

north-east coast of Brazil. The presence of this boundary was confirmed, and when the rocks on either side of the predicted boundary were dated they were found to be exactly as expected: 2000 Ma to the west and 600 Ma to the east. Apparently a fragment of the 2000 Ma West African province had been left on the South American continent at the time of the break-up.

Tuzo Wilson also used radiometric dating information to confirm the fit of North America and Europe across the North Atlantic. He also showed that all oceanic islands are less than 150 Ma old, apart from those that were obviously continental fragments (such as Madagascar and the Seychelles in the Indian Ocean and the Falklands in the South Atlantic), which supported the idea that the ocean basins were youthful features. More detailed work on the ages of islands in the South Atlantic indicated that the rate of movement away from the mid-oceanic ridge is somewhere between 2 and 6 cm a^{-1}.

Surprisingly, these developments were not sufficient to sway the majority of the geological community, who remained unconvinced and markedly antipathetic to the whole idea of lithospheric mobility. Most geophysicists still maintained that the Earth was too strong to allow convection currents in the mantle. Many geologists ignored the palaeomagnetic results and maintained that nothing new had been produced. All this kind of evidence had been rejected in Wegener's time and they saw no reason to reverse that decision. As more and more palaeomagnetic results accumulated, it became increasingly awkward for them to maintain this stance.

There were at this time considerable numbers of marine geologists, particularly in the USA, who recognised that something fundamental was going on around the mid-oceanic ridges, but who were almost unanimous in saying that it was not sea-floor spreading. The reasons for this were that the model was too simplistic and that the marine geologists had too many facts; they could not see the wood for the trees!

The revolution that changed the highly speculative idea of sea-floor spreading into today's substantive model of plate tectonics hinged on developments from the basic idea of sea-floor spreading, which caused this group of marine geologists to change their minds. Once they were converted the revolution was under way and the rest of the geological world followed, with very few exceptions. These crucial developments will form the topic of the next chapter.

CHAPTER SIX

The revolution

Palaeomagnetism initiated the revolutionary breakthrough that occurred next, but this time it was magnetic reversals that were to provide the key. As explained in Chapter 3, the direction of the magnetism in some rocks is the complete opposite of that expected and, until this time, reversals had been regarded as no more than a nuisance in studies of polar wandering and continental drift.

The main problem initially was whether these reversals in rocks were self-reversals, owing to certain minerals having the capability of assuming remanent magnetism in the opposite direction to the applied field, or a record of real reversals of the Earth's magnetic field, when the magnetic North Pole coincides with the geographical South Pole. Fortunately it was possible to distinguish between these two phenomena because self-reversals, being mineralogically determined, could only be of local importance, whereas the effect of true field reversals would be world-wide.

The concept of self-reversals was supported by a considerable amount of both theoretical and experimental work, which showed that certain minerals do behave in this way when cooled through their Curie point. There was also one well publicised natural example of self-reversal in a lava flow from Japan. The evidence supporting field reversals, however, was equally strong; Figure 6.1 is typical of many recorded examples and shows the effect of an igneous rock intruding sedimentary rocks at a time of reversed field. The sedimentary rocks have normal magnetism except for that part adjacent to the igneous rock, where the magnetism has been reversed. Such a distribution must have been acquired when the minerals of the igneous rock and the adjacent parts of the sedimentary rock, heated by the intrusion, cooled through their Curie point and acquired the ambient reversed field.

Previous work had shown that Pliocene and Pleistocene lava flows often contained good examples of magnetic reversals, so very detailed investigations were begun on such lavas in Iceland and northwestern USA. These established that reversals were unrelated to mineralogy but were dependent on the age of the rock. In other words, self-reversal was a very rare phenomenon, and it could be assumed that practically all reversals recorded

92

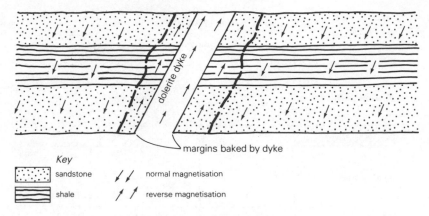

Key

☐ sandstone / / normal magnetisation

☰ shale ↗ ↗ reverse magnetisation

Figure 6.1 Evidence of reversals of the Earth's magnetic field.

in rocks were the result of true field reversals. This meant that a sequence of reversals recorded in a particular group of rocks would have world-wide applicability, provided the rocks could be dated with sufficient accuracy, as apart from dating there was no way to distinguish one reversal from another. Dating, however, proved to be a matter of considerable difficulty at that time.

Basalts were generally dated by the standard K/Ar method using the isotope dilution technique (see Ch. 3). However, when this method was applied to rocks less than about 10 Ma old the results were often inconsistent, because the ^{40}Ar being measured was contaminated by extraneous argon. This contaminating argon was found to have been absorbed by the glass apparatus during previous argon determinations, being released when the glass was heated in subsequent determinations. The amount involved was extremely small and was of no significance when a relatively large amount of ^{40}Ar (from minerals of great age) was being determined. However, when the amount of ^{40}Ar was very small, as it was with young rocks, the contamination was sufficient to make the results meaningless. In 1956 an all-glass spectrometer was developed at the University of California that could be heated while under vacuum before each measurement to drive off the absorbed argon. Once decontaminated in this way the spectrometer could measure minute amounts of radiogenic argon very accurately and so the ages of Pleistocene and Pliocene rocks could be radiometrically determined with considerable accuracy.

By 1963 sufficient data had been amassed to establish that field reversals were world-wide synchronous events and this permitted a tentative time-scale to be published (Fig. 6.2a). Tentative as this was, it was sufficient to initiate a major revolution in the Earth sciences, for it enabled Vine and Matthews (from Bullard's department at Cambridge) to combine the observed mid-oceanic ridge magnetics with the concept of sea-floor spreading (discussed in Chs 4 & 5).

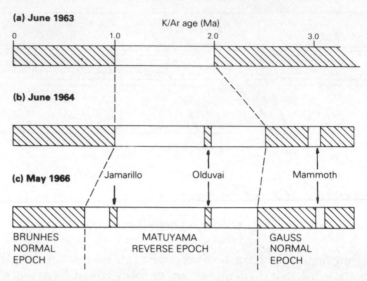

Figure 6.2 The development of the magnetic reversal timescale.

Vine and Matthews' work showed that the Carlsberg Ridge, in the north-west Indian Ocean, was composed of basaltic lavas and topographically consisted of a central rift valley flanked by high submarine scarps. The topography seemed to be related to the magnetic anomaly pattern, for a central negative anomaly coincided closely with the rift valley and positive anomalies with the flanking scarps. However, this relationship was shown to be more apparent than real, when two submarine volcanoes of similar shape were found to have anomalies of opposite sign. Calculations showed that this difference could be accounted for if one of the volcanic features was normally magnetised and the other reversely magnetised, suggesting that the ocean floor could be either normally or reversely magnetised. This was tested by comparing observed marine magnetic profiles with profiles computed assuming that the ocean floor was composed of blocks of basalt that were alternately normally and reversely magnetised (Fig. 6.3). The results were in good agreement, suggesting that 50% of the ocean floor could be reversely magnetised.

Vine and Matthews explained these results by combining the concepts of sea-floor spreading with the periodic reversals of the Earth's magnetic field. If new ocean floor is formed where basalt wells up at the centre of an oceanic ridge, it will be magnetised in the prevailing direction of the Earth's magnetic field. As sea-floor spreading continues, this material will split down its centre and half will move laterally to each side of the ridge. The split will be filled by new basalt to be magnetised in its turn as it cools below the Curie point. At some time in this process, the Earth's field may undergo one of its periodic reversals and this will cause subsequent basalts to be magnetised in the opposite direction. A number of such reversals occurring

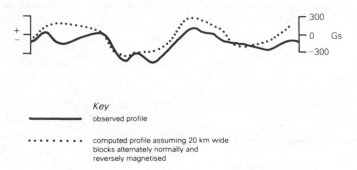

Key

─────────── observed profile

• • • • • • • • • computed profile assuming 20 km wide
blocks alternately normally and
reversely magnetised

Figure 6.3 Magnetic anomalies of the Carlsberg Ridge.

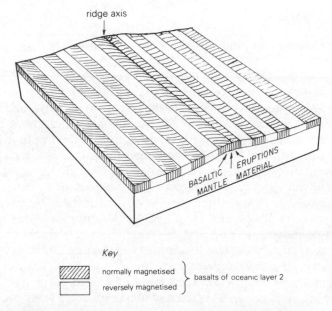

Key

normally magnetised ⎱
                    ⎰ basalts of oceanic layer 2
reversely magnetised

Figure 6.4 The origin of linear magnetic anomalies.

during sea-floor spreading would result in the development of successive strips of the ocean floor that would be alternately normally and reversely magnetised and symmetrically placed on either side of and parallel to the ridge (Fig. 6.4). This was a simple and plausible explanation of the linear magnetic anomalies of the eastern Pacific (see Ch. 4) and it did not require any major abrupt lateral change in the mineralogy of the rocks of the oceanic lithosphere. As was subsequently realised, this was the fundamental breakthrough to the plate tectonic model, but at the time most marine geologists rejected the idea on the grounds that there was little concrete evidence to support such speculations. Furthermore the model proposed was too simple; it was felt that nothing in geology could be quite so simple.

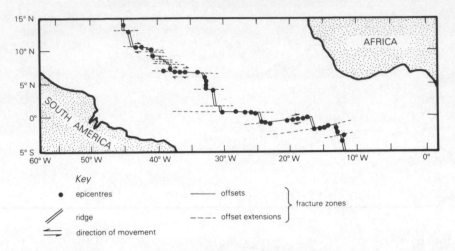

Key

• epicentres ——— offsets

⟍⟍ ridge – – – – offset extensions

⇆ direction of movement

⎫
⎬ fracture zones
⎭

Figure 6.5 Offsets of the Mid-Atlantic Ridge and distribution of earthquake epicentres.

Vine was as aware as anybody else of the speculative nature of this work and the need for more concrete data. These were provided by Tuzo Wilson's ideas concerning the movement of the sea floor on and about the mid-oceanic ridges, which derived from investigations showing that both the East Pacific Rise and the Mid-Atlantic Ridge, instead of being continuous features, were offset for distances of up to 1000 km along fracture zones (Fig. 6.5). Conventional wisdom interpreted this pattern to mean that the ridge had formed first and then been disrupted (offset) by transcurrent faulting. If movement was still taking place, the opposite sides of such a fault would be moving in opposed directions and earthquakes would be generated along its entire length (Fig. 6.6). However, this presented a major problem, for it had already been found that earthquake epicentres were confined to the actual ridge crest and to that part of the fracture zone between the offset ridge crests. What had to be explained was how one part of the fracture zone was seismically active and yet its direct continuation, outside of the ridge–ridge offsets, was not (Figs 6.7 & 6.5). In mid-1965 Tuzo Wilson suggested that this paradox could be resolved if the ridge–ridge offsets were examples of a previously unrecognised type of fault in which

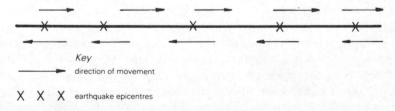

Key

——▶ direction of movement

X X X earthquake epicentres

Figure 6.6 Transcurrent faults and earthquake epicentres.

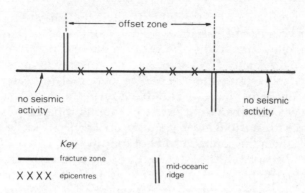

Figure 6.7 Earthquake epicentres on fracture zones.

the movement was horizontal but different from that occurring along transcurrent faults and which he named **transform faults**.

If all of the sea floor is moving away from the mid-oceanic ridge, then it follows that across the ridge–ridge offset (Fig. 6.8) sea-floor movements are opposed and hence earthquakes are generated. Outside of the offset zone, the sea floor on either side of the fracture zone is moving at the same speed and in the same direction away from the ridge and hence no earthquakes will be generated.

This concept gained strong support from its ability to explain a number of puzzling facts. The immense escarpments that occur along some portions of these fracture zones and then disappear laterally, as had been reported from the north-east Pacific (see Ch. 4) were now explicable. The mid-oceanic ridges are topographically high because hot basalt is emplaced

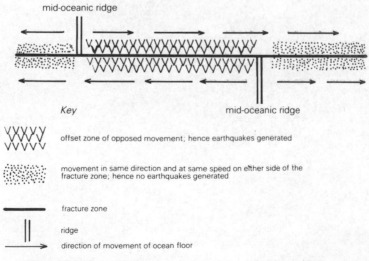

Figure 6.8 Tuzo Wilson's solution – the transform fault.

97

during the spreading process, while the abyssal plains are topographically much lower because the basalt cools and contracts as it moves away from the ridge. It follows that ridge offsets cause the sea floor in contact across a fracture zone (at A and B in Fig. 6.9) to be of different ages and hence at different distances from the ridge. The part that is nearer (A) will be topographically higher than the part that is further away (B) and an aseismic escarpment will occur between. Such topographic differences will gradually disappear as both A and B move laterally, cool and subside, and eventually A and B will reach the common level of the abyssal plain.

The general shape of the Mid-Atlantic Ridge is close to the shape of the rifted margins of the flanking continents and so this shape could be explained in terms of the necessary pattern of ridge fragments and linking transform faults needed to follow the initial rift between the Americas and Africa and Europe (Fig. 6.5). No transcurrent movement on fracture zones is then required.

The San Andreas Fault of California was also re-interpreted as a transform fault, rather than a transcurrent fault, connecting a ridge segment in the Gulf of California with one off the coast of British Columbia (Fig. 6.10). It was this last ridge fragment, called the Juan de Fuca Ridge, that attracted Vine's attention, for what he was lacking were magnetic anomaly data across an ocean ridge on an areal basis rather than the isolated profiles previously available. The Juan de Fuca area had been covered by an extension of the magnetic anomaly survey of the north-east Pacific (see Ch. 4) and Tuzo Wilson had identified this as an active oceanic ridge

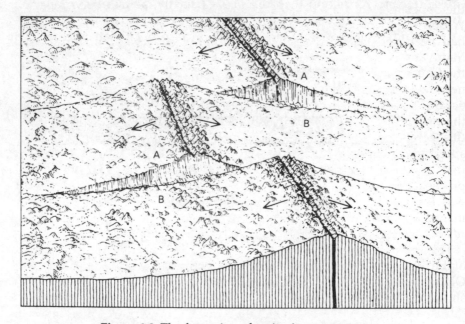

Figure 6.9 The formation of aseismic escarpments.

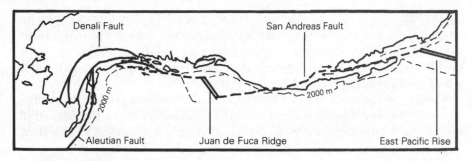

Figure 6.10 Re-interpretation of the San Andreas Fault as a transform fault.

fragment. It was obvious that the anomaly pattern was highly symmetrical with regard to the ridge (Fig. 6.11), which was just what Vine required, and the result of his joint investigation with Tuzo Wilson was published in 1965.

By this time a better understanding of the problems involved enabled profiles to be computed in a more realistic manner than was possible in 1963 when Vine and Matthews published their original paper. It had now been established that the magnetism giving rise to the anomaly pattern was mostly confined to the oceanic layer 2 basalt. In addition the timetable of magnetic reversals had been considerably elaborated (Fig. 6.2b). It was now apparent that reversals did not occur at regular time intervals but were quite irregular in their occurrence. Furthermore, within each major magnetic epoch there were short periods, or **events**, during which the polarity was the opposite to that of a particular epoch. By the time that Vine and Tuzo Wilson were working on the Juan de Fuca Ridge, just two of these events had been recognised; the Olduvai normal event at 1.9 Ma during the Matuyama reverse epoch and the Mammoth reverse event at 3.0 Ma within the Gauss normal epoch. When model profiles were computed, using this new reversal timescale and with the anomaly source rocks confined to oceanic layer 2, there was good agreement with the observed profile in that the general pattern of positive anomalies matched (Fig. 6.12). However, this agreement was only achieved by abandoning the basic idea that spreading from the ridge was a uniform process and postulating instead that spreading had been faster prior to 2 Ma ago and had since then proceeded at a slower rate (Fig. 6.12b). This was not very satisfactory, for it was felt that the continental rocks reversal timescale and the sea-floor magnetic anomaly pattern should fit together without having to use such an *ad hoc* correction.

The interpretation of an irregular spreading rate hinged on the correlation of the first normal event prior to the present Brunhes normal epoch with the Olduvai event at 1.9 Ma, but in the following year this correlation was shown to be incorrect. Additional data from continental lava flows and intrusives revealed that the boundary between the Brunhes normal and Matuyama reverse epochs (Fig. 6.2c) was at 0.7 Ma rather than 1 Ma and that the first event beyond this point was in fact the Jamarillo at about 1 Ma, which had not been identified before, and not the 1.9 Ma Olduvai event. When this revised timetable was used to compute the magnetic anomaly profile, it was found to match the observed profile extremely well, assuming a uniform spreading rate of 3.2 cm a^{-1} (Fig. 6.12).

Almost immediately the production of the Eltanin-19 profile, across the crest of the East Pacific Rise, gave an immense boost to the idea that symmetrical spreading from mid-oceanic ridges is reflected in sea-floor

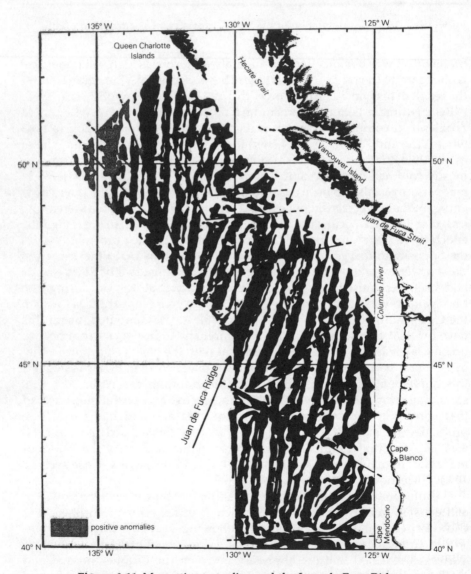

Figure 6.11 Magnetic anomalies and the Juan de Fuca Ridge.

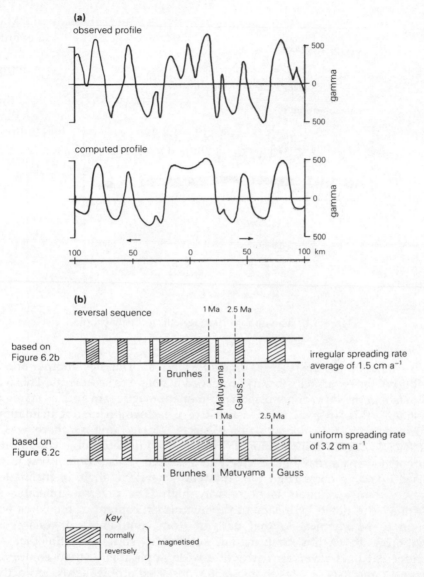

Figure 6.12 Magnetic anomaly pattern and spreading rate (Juan de Fuca Ridge).

magnetic anomaly patterns. The symmetry of this profile was so evident that it could not be disputed and this caused people who had been bitterly opposed to the whole idea to think that Vine and Matthews might be right after all. This became a much stronger possibility when the computed profile making use of the new reversal timescale (Fig. 6.2c) was found to fit almost perfectly with the observed profile at a uniform spreading rate of 4 cm a^{-1} away from the ridge (Fig. 6.13).

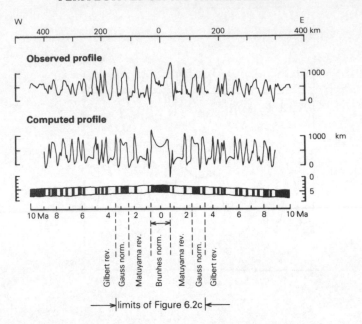

Figure 6.13 The Eltanin-19 magnetic anomaly profile.

The reliability of the reversal timetable was confirmed by another line of evidence, the reversal history as preserved in deep-sea sediments. This was of course in the form of depositional remanent magnetism and, as stated in Chapter 3, this can give a reliable measure of declination but not inclination. However, this was not a major obstacle, for as long as there was a reasonable degree of inclination it could be used to determine whether the sediment was normally or reversely magnetised. This problem was minimised by using cores from the Antarctic where the angle of inclination preserved in sediments is necessarily high. The reversal timetable so obtained and dated by means of the microfossil content of the cores was found to be identical to that derived from continental igneous rocks (Fig. 6.2c). If, as this confirmation seemed to show, the timetable of continental field reversals had world-wide applicability, then computed magnetic anomaly profiles should match observed profiles across any active ridge area anywhere in the world, provided that account was taken of the possibility that rates of spreading could differ from one ridge area to another. That this was indeed so was demonstrated when the reversal timetable was used to compute profiles across the mid-oceanic ridge in both the North and South Atlantic and these were found to equate almost exactly with the observed profiles (Fig. 6.14).

For most geologists these developments established the reality of sea-floor spreading, but yet further confirmation was almost immediately provided by the seismologists. They were now receiving much more

102

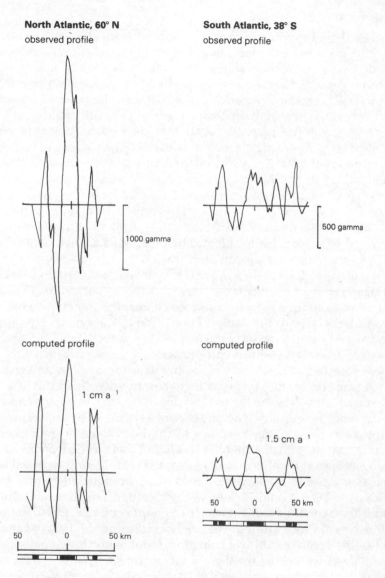

Figure 6.14 Atlantic magnetic anomaly profiles.

detailed seismograph records from a world-wide standardised seismograph network of more than 150 stations. By recording both body and surface waves, this network furnished data of greater sensitivity, greater reliability and broader geographical coverage than had been available before. This allowed very detailed studies to be made of the movements responsible for the generation of earthquakes on or about the mid-oceanic ridge, such movements having been shown earlier to be confined to the ridge itself and to the ridge–ridge offset zones (Figs 6.5 & 6.7). These better data showed that earthquakes located on the ridge crest were characterised by normal faulting resulting from tension acting at right angles to the ridge and that earthquakes on fracture zones were due to horizontal movement in the directions predicted for transform faults (Figs 6.5 & 6.8). These results not only confirmed the transform nature of the faulting at mid-oceanic ridges but in so doing strongly supported the idea of sea-floor spreading.

As well as providing confirmation that the lithosphere was being formed at the ridges, seismologists were able to provide evidence of its destruction elsewhere. In Chapter 3 the distribution of deep-focus earthquakes in Benioff Zones, which dip under the oceanic trenches and adjacent island arcs or fold mountains, was discussed. One of the major paradoxes to come out of this work was that these deep-focus earthquakes well within the mantle were the result of rock fracture, where on other geophysical grounds it would be reasonable to suppose that materials at such depths would be plastic and hence not yield by fracture.

Detailed work on seismic waves from deep-focus earthquakes below the Tonga Trench showed that the P and S waves generated were transmitted much more effectively to stations located near the trench, where the waves would pass close to the Benioff Zone, than to stations where these waves had to pass through areas of the mantle (Fig. 6.15). Thus the P and S waves received from a deep-focus earthquake at Tonga were much stronger than those received at Fiji. This was explicable if the material traversed by the waves en route to Tonga had greater rigidity than did the material traversed between the focus and Fiji. On this basis the results were interpreted as indicating the presence of a slab of low-temperature material projecting down along the trace of the Benioff Zone into more normal mantle material. Furthermore, it was found that strong P and S waves were also received at Raratonga, 1800 km to the east, having travelled by way of the thin oceanic lithosphere, so it could be concluded that there was continuity between this section of the oceanic lithosphere and the material along the Benioff Zone. This was strong support for the idea that the oceanic lithosphere, which formed at the mid-oceanic ridge and moved away from it across the ocean basins, was being thrust down into the mantle. This would place deep within the mantle a lithospheric slab, which could fail under stress to yield the seismic pattern that defines the Benioff Zone.

This accumulation of confirmatory data caused an almost complete change of opinion in the ranks of marine geologists in less than a year, between mid-1966 and early 1967. Whereas before this time only a few believed in sea-floor spreading, afterwards very few did not. This change was spearheaded by scientists at the Lamont Geological Observatory, who were responsible for both the Eltanin-19 profile and the analysis of the seismic data at the ridges and trenches discussed above. They were also responsible for a great deal of the work involved in the next phase in which the implications of sea-floor spreading were evaluated, for they possessed a

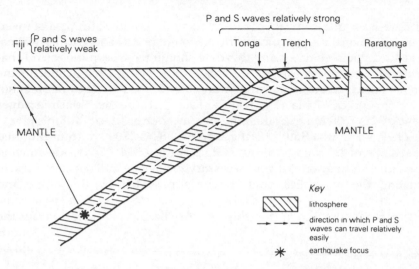

Figure 6.15 The lithosphere and Benioff Zone, and the transmission of seismic waves.

great quantity of magnetic anomaly data from all the oceans of the world. Once the key had been provided by the Eltanin-19 profile, the way was open for it to be deciphered.

Investigations showed that magnetic anomaly patterns were bilaterally symmetrical about mid-oceanic ridges for thousands of kilometres across the world's ocean basins. In addition it was recognised that every anomaly profile contained a set of common characteristic peaks. To aid discussion, these anomaly peaks, or anomalies for short, were numbered 1 to 32, from the ridge crests outwards. This enabled profiles to be correlated from one ocean basin to another, given that each anomaly was of the same age in all basins (Fig. 6.16a), having formed as a result of sea-floor spreading during a particular reversal period. It was found that all profiles from a mid-oceanic ridge to anomaly 32 could be simulated by computations that required 171 reversals of the Earth's field during this time interval (Fig. 6.16a).

Obviously the trace of a particular numbered anomaly on any ocean floor represented the same time period. This correlation was, for the most part, purely relative, as only the three anomalies nearest the ridge fell within the 4 Ma timespan covered by the radiometrically determined continental reversal timescale (Fig. 6.2). Thus only 100–200 km of the ocean floor, out of the 1800–3100 km involved in the magnetic anomaly pattern, was directly dated, for the oceanic layer 2 basalts that contain the magnetic anomaly were not available for direct radiometric dating. Therefore the next major problem was to extend this radiometrically based timescale from anomaly 3 out to anomaly 32.

From previous work it was known that the spreading rate from any given active ridge was uniform over the past 4 Ma. As the spreading rate before this time was unknown, it was decided to assume uniform rates for all oceans. Knowing the distance from the ridge and the rate of spreading, anomaly 32 could be dated and this date should have been the same for all oceans. However, this was not the case. In the North Pacific, which has a current spreading rate of 3.2 cm a^{-1}, anomaly 32 is 3100 km from the ridge, giving an age of 97 Ma (Fig. 6.16 & Table 6.1). In the South Atlantic, anomaly 32 is 1500 km from the ridge and at a rate of 1.9 cm a^{-1} this gives a date of 79 Ma. In the South Pacific, anomaly 32 is 1800 km from the ridge and at a rate of 4.0 cm a^{-1} this gives it an age of 45 Ma. To decide between these different timescales it was necessary to have an independent method of dating the anomalies, and this was provided by the ocean-floor sediments.

Figure 6.17 shows the relationship between the age of sea-floor sediments

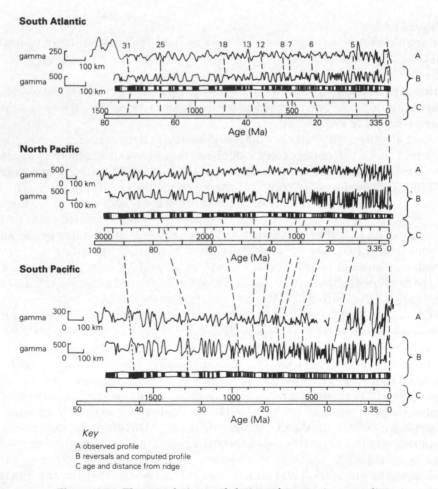

Key
A observed profile
B reversals and computed profile
C age and distance from ridge

Figure 6.16 The correlation and dating of magnetic anomalies.

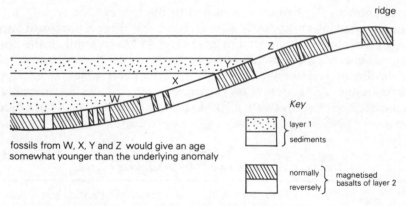

fossils from W, X, Y and Z would give an age
somewhat younger than the underlying anomaly

Figure 6.17 Relationship between sediments and magnetic anomalies.

(layer 1) and the underlying magnetic anomalies (layer 2). From the concept of sea-floor spreading it follows that the basal sediment of layer 1 was deposited after the basalts of layer 2 were formed at each location, so that the sediment is younger in age than the immediately underlying layer 2 anomaly, but not all that younger by geological standards. If such sediments could be dated, they would give a reasonable measure of the age of the underlying magnetic anomaly.

Table 6.1
The age of anomaly 32.

	South Atlantic	North Pacific	South Pacific
spreading rate for past 4 Ma (cm a^{-1})	1.9	3.2	4.0
distance of anomaly 32 from ridge (km)	1500	3100	1800
age of anomaly 32 by extrapolation (Ma)	79	97	45

However, in 1966 only three sediment samples from such locations were available and, as the only means of dating them was by their contained fossils, an approximate date was all that was possible. In the north-east Pacific, off the coast of British Columbia, a near-basal sediment of layer 1 had been given a fossil age of Lower Miocene to Middle Miocene (13–26 Ma) and this sample occurred over anomaly 6 (Table 6.2). The North Pacific and South Atlantic anomaly timescales gave anomaly 6 an age of 22 and 20 Ma respectively, while from the South Pacific the age was only 14 Ma (Table 6.2). Another basal sediment from the western North Pacific, located just west of anomaly 32, had been assigned to the Upper Cretaceous (70 Ma) while, as stated above, anomaly 32 was dated at 97 Ma on the North Pacific scale, 79 Ma on the South Atlantic scale and 45 Ma in terms of the South Pacific scale. From the South Atlantic, towards the South American coast, a basal sediment just west of anomaly 31 had been assigned an Upper Cretaceous age (70 Ma) while, on the three oceanic timescales, anomaly 31 is dated at 89 Ma (North Pacific), 73 Ma (South Atlantic) and 40 Ma (South Pacific).

Table 6.2
Dating anomalies of layer 2 by basal layer 1 fossils.

	North-east Pacific	Western North Pacific	South Atlantic
anomaly to be dated	6	32+	31+
fossil age (Ma)	13–26	70 (approx.)	70 (approx.)
age (Ma) according to			
North Pacific scale	22	97	89
South Atlantic scale	20	79	73
South Pacific scale	14	45	40
(see Table 6.1)			

These sparse data indicated that there was something very aberrant about the South Pacific timescale, and even though there was greater similarity between the other two they still differed considerably. In the end the South Atlantic timescale was preferred to that of the North Pacific, because spreading from the latter may have been affected when the Pacific Ridge was overwhelmed by the westward movement of North America, whereas there appeared to be no such disturbance in the case of the South Atlantic. It was then possible to use the South Atlantic timescale to extend the radiometric reversal timescale back through the recorded 171 reversals and to relate them to the timescale developed in Chapter 3 (Fig. 6.18).

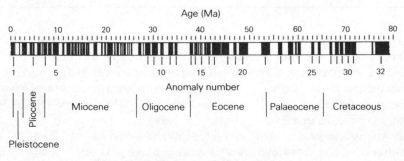

Figure 6.18 The extended reversal timescale.

When the choice was made to use the South Atlantic as the model for sea-floor spreading, it was merely the best available out of three rather poor alternatives, and the possibilities of error in using this extrapolation were particularly stressed by the Lamont scientists who did this work. In retrospect, it is amazing how little the original estimates have been amended by a considerable body of later data.

Confirmation of the suggested timescale came from the results of a programme initiated in 1968 by various American institutes with the acronym JOIDES (Joint Oceanic Institutes for Deep Earth Sampling). JOIDES undertook the systematic investigation of all ocean basins by deep-sea drilling and core sampling from the specially constructed ship, Glomar Challenger, using the techniques for drilling under several kilometres of water developed during the Mohole project (see Ch. 3).

Typical of the JOIDES results were those obtained from the South Atlantic in 1969, where a series of nine holes was drilled across the mid-oceanic ridge. All holes penetrated the entire thickness of layer 1 and entered the basalt of layer 2. The age of the base of layer 1 was determined by its microfossil content and applied to the layer 2 basalts beneath. Direct dating of layer 2 was not possible because the basalt of layer 2 had been altered by sea water, so that K/Ar dates could not be relied upon. It was not possible to make use of the magnetic reversals of layer 1, as cores were retrieved only intermittently from any given drill hole. When the microfossil ages were plotted against their distance from the ridge axis, the result was an almost straight-line relationship (Fig. 6.19) in close agreement with the predicted ages. This confirmed the previous extremely tentative extrapolation that, for the last 80 Ma, the South Atlantic seafloor has been moving away from the ridge at a remarkably uniform 1.9 cm a^{-1}.

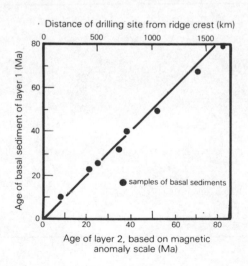

Figure 6.19 Sediment age and magnetic anomaly age in the South Atlantic.

109

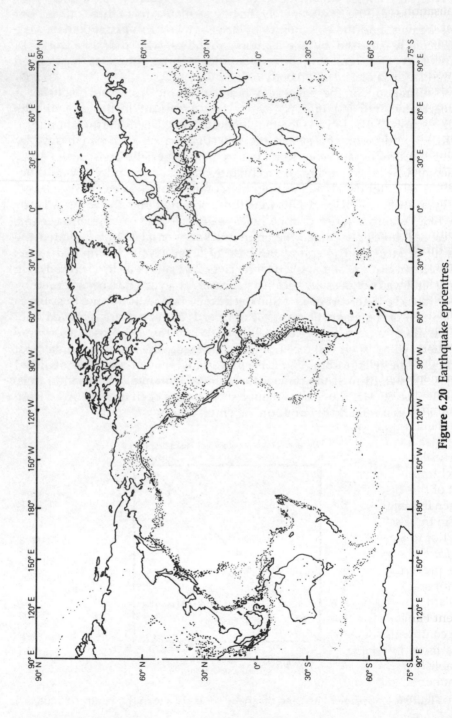

Figure 6.20 Earthquake epicentres.

The final rationalisation of the plate tectonics model stemmed from the realisation that the ocean-wide preservation of magnetic anomalies meant that nothing happened to this lithospheric material from the time it was formed at the mid-oceanic ridge to the time of its destruction by subduction. In other words all the action, such as volcanism and earthquakes, is concentrated in narrow zones where the lithosphere is either being created or destroyed. On a world-wide basis the best way to locate these narrow zones was by plotting the distribution of earthquake epicentres, in the same way as was done initially to plot the course of the mid-oceanic ridge (Fig. 4.8). When this was done, it became apparent (Fig. 6.20) that this defined a set of rigid lithospheric plates covering the entire surface of the Earth (Fig. 6.21). It was possible to determine the relative motion of these plates from the first movements caused by the earthquakes concentrated at their boundaries, which enabled three different kinds of boundaries to be recognised (Fig. 6.22): diverging boundaries associated with the tensional faulting of the mid-oceanic ridge; converging boundaries associated with the thrust faulting below oceanic trenches and continental collisions; and conservative boundaries associated with transform faulting.

An equally important development was the realisation that the rigidity of the lithospheric plates made it possible to plot their movements with considerable accuracy, for what was involved was a problem in spherical geometry, developed by Euler some 200 years before. This had already been used by Bullard in his computer fit of the Atlantic continents (Fig. 5.5). The new developments showed that the method was not just restricted to continents but had world-wide applicability. In addition, the whole task was made much easier by the great amount of new data available from the ocean basins. Thus the establishment of an ocean basin timescale based on magnetic anomalies meant that each anomaly represents a particular time in the past, so that lines of equal age (isochrons) can be drawn coinciding with particular anomalies (Figure 6.23 is an early attempt); a succession of isochrons is then a trace of plate movement on the ocean floor. The working out of this history has been a main centre of activity in Earth sciences, but when it is completed it will only go back some 80 Ma, the limit of the ocean basin timescale.

Prior to 80 Ma the ocean basin evidence peters out and there is no *direct evidence* to say how long plate tectonics has been operative. It was seen in the previous chapter how the polar wander curves, determined from continental rocks, could be combined with the exact fitting techniques of Bullard to follow continents from the time they formed part of the supercontinent Pangaea (some 200 Ma ago in the early Triassic) through the following continental dispersal. Prior to 1968, this dispersal only had the support of a minority of Earth scientiests. After 1968, with plate tectonics firmly established, there were no longer any doubts that the proposed history of continental movement was generally correct. However, the number of possible initial reassemblies suggested by various authors, particularly of

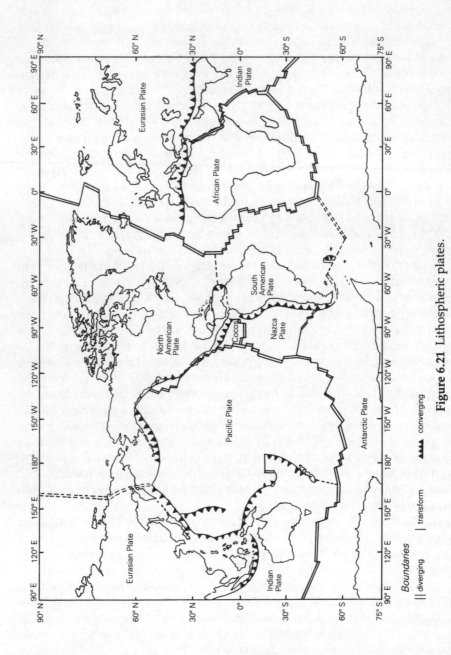

Figure 6.21 Lithospheric plates.

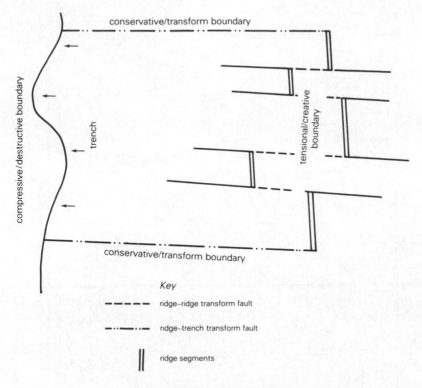

Figure 6.22 Plate boundaries.

Gondwanaland (Fig. 6.24), emphasises the equivocal nature of the evidence being used, but there was no doubt that the causative mechanism of continental movement was plate tectonics.

As plate tectonics was involved in the break-up of Pangaea, it would be reasonable to suppose that the same process was involved in its assembly, and there are various lines of evidence to support this contention. Thus the present-day Appalachian fold belt marks the site of a mountain range created during the formation of Pangaea, and among the folded rocks are some that consist of fragments of the three layers of the oceanic lithosphere: cherts (from sediments), basalts and ultrabasic rocks, which together are given the name **ophiolites**. Their presence implies that the original Appalachian Mountains were created at a plate edge probably as a result of a continental collision, before which an ocean was eliminated by the operation of plate tectonic subduction processes. This interpretation gains further support from the presence of glacial deposits, which indicate that a south polar ice cap covered the Sahara 400 Ma ago. At the same time eastern North America lay near the Equator (Fig. 6.25). The close proximity of Pole and Equator in Pangaea reconstructions can only be reconciled by Africa and North America being separated, 400 Ma ago, by an ocean some

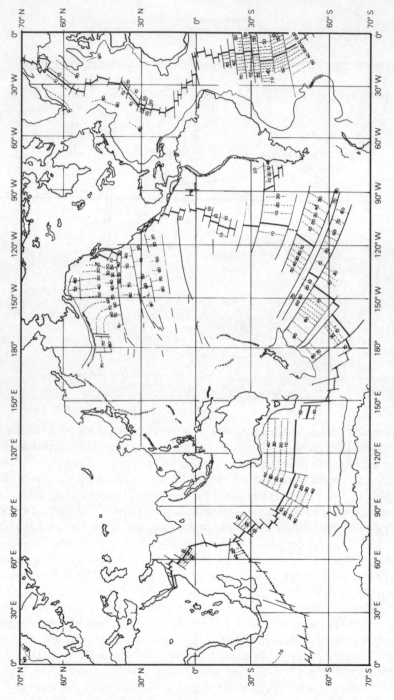

Figure 6.23 Isochron map (Ma) of the ocean floor.

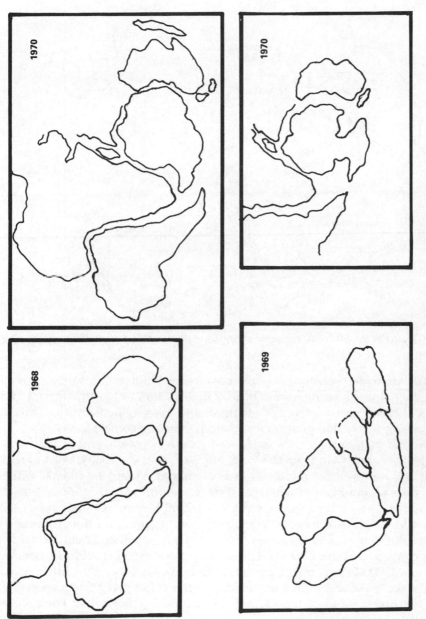

Figure 6.24 Variations of Gondwanaland reassembly, 1968–70.

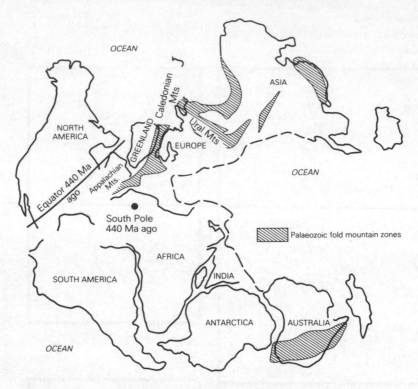

Figure 6.25 Evidence for plate tectonics within a reconstructed Pangaea.

10 000 km wide. Subsequent plate movement eliminated this ocean, creating the Appalachian mountain belt by the collision of North America and Africa, in the process of forming Pangaea. By using evidence such as this, it is possible to say that plate tectonics was operating some 600 Ma ago.

Beyond that, however, the evidence becomes somewhat more equivocal. Ophiolites are absent from the long, narrow, intensely folded belts of rock that are the erosional remnants of Precambrian mountain chains, which casts some doubts on whether they were formed at plate margins. However, it is only in rocks older than 2500 Ma that doubts about the operation of plate tectonics become really pronounced, for the proportions of the different rock types appear to alter in a fundamental manner, which has convinced some Earth scientists that plate tectonics did not operate prior to 2500 Ma ago, but at present nothing has been proved either way.

The consequences of this revolution are that it has given Earth scientists, for the first time since the idea of the contracting Earth had been abandoned, a broad-based integrative view of their subject, which has resulted in an immense revivification of the whole of the Earth sciences, not only by the direct application of plate tectonics to many previously moribund areas but also in making more acceptable broad-scale approaches in many areas previously submerged in the minutiae of detailed studies. Nowhere can this be seen more clearly than in the recent attempts that have been made to explain the Earth's major topographic features, which is the subject of the next chapter.

CHAPTER SEVEN

The Earth's macrotopography in terms of plate tectonics

Towards the end of the first chapter an attempt was made to explain the development of the Earth's major surface features in terms of it being a cooling, contracting body. In this last chapter the Earth's macrotopography will once again be discussed, but this time in terms of plate tectonics. This, of necessity, will be highly speculative, for such a discussion will involve topics that are in the forefront of present-day research.

From the plate tectonic point of view, the relationships between processes and topographic form are much clearer within the ocean basins than on the continents. The main reason for this lies in the compositional difference between the two; the basaltic nature of the ocean basin rocks compared to the more silica-rich continental rocks.

In terms of density, this means that there is little or no difference between the ocean basin rocks and the underlying mantle, so that subduction can easily occur. The silica-rich materials of the continents are, however, decidedly less dense, so that it is very difficult, if not impossible, for them to be subducted. The ease of subduction within ocean basins means that most topographic features are the product of the present plate tectonic cycle, so that the relationship between process and topography is relatively simple and straightforward. In contrast, the continental materials remain at the surface to be reworked again and again during many cycles of plate tectonic activity, which means that in general continental topography is polycyclic and thus much more difficult to interpret in terms of plate tectonic processes.

Another consequence of this compositional difference is that continental rocks deform at a lower temperature than do those of the ocean basins, so that it is more likely that they will react to plate motion by deformation, whereas those of the ocean basins will tend to fracture. Hence plate boundaries are generally narrow and well defined within ocean basins and become broad and diffuse on the continents, so that the relationship between process and form in the latter case is much more obscure.

In addition to these compositionally related factors of density and ductility, there is yet another factor involved, that of weathering. Once again, as it has such a great effect on land areas compared to sub-oceanic environments, it makes the relationship between topography and plate tectonics that much more difficult to establish on the continents.

Therefore, ocean basin topography is considered first, before approaching the much more difficult problems posed by continental topography.

The Pacific Ocean provides the simplest and most straightforward example of topography related to plate tectonics, for here there are developed mid-oceanic ridges, abyssal plains, peripheral trenches and their associated island arcs, all of which are explicable in terms of a plate being created at a diverging margin, moved across an ocean basin and subducted at a converging margin. The topographic difference between the mid-oceanic ridge and abyssal plain is related to the cooling and contraction of the lithospheric plate as it moves away from its point of origin and is such that sea-floor depth is directly related to the square root of its age. This relationship also explains why the Mid-Atlantic Ridge is such a topographically marked feature compared to the East Pacific Rise, for its lower spreading rate means that age, and hence sea-floor depth, increases at a much greater rate away from the ridge than is the case for the faster-spreading East Pacific Rise. Trenches are the surface reaction to the downward plunge of the dense cold lithospheric plate into the mantle, which carries with it water-saturated ocean-floor sediments. These, together with the somewhat hydrated uppermost part of the plate, react to the higher temperatures encountered to produce a water-rich melt. The melt penetrates upwards through the mantle of the non-subducting plate, in the process melting certain of its minerals and eventually reaching the surface as lava, pumice or ash, to form the volcanic island arcs, which are aligned parallel to the trenches and directly above the point where, at a depth of about 100 km, the downgoing plate departs from the overlying plate.

As well as the island arcs running parallel to the trenches, the Pacific is characterised by linear chains of islands that approach the trenches at a high angle. The Hawaiian chain, which extends across the central Pacific from ESE to WNW over a distance of some 3000 km, is probably the best known example. Along these chains there is a progression from active volcanic islands at the end nearer the mid-oceanic ridge (ESE in the case of the Hawaiian chain) through islands where volcanism has ceased and which have fringing reefs and then barrier reefs, to the final situation where there is an atoll with no central island visible, at the end nearer to the trenches. This lateral change of form is of course the same as that envisaged by Darwin (see Ch. 4) and his concept was finally vindicated by plate tectonics, for the cooling and contraction of the lithospheric plate as it moves away from the mid-oceanic ridge provides an adequate mechanism for the gradual submergence of a volcanic island; this submergence had previously been the stumbling block to the general acceptance of his theories. At the same time guyots (see Ch. 4) were seen as being volcanoes that when active had never reached the near-surface zone, where corals flourish, and therefore had left no surface evidence of their presence in the form of reefs and atolls.

However, these features also provide something of a mystery in plate tectonic terms, for the volcanic eruptions leading to the formation of these island chains occur in the middle of plates well away from any boundaries where action is supposedly concentrated. Recent geophysical work has suggested that such volcanism is associated with the upwelling limbs of convection currents in the mantle.

From this discussion it is apparent that many of the macrofeatures of the ocean basins can be explained relatively simply in terms of plate tectonics. In recent years, however, intensive investigations on and about the mid-oceanic ridges have detected a range of features that are not so easily accounted for. The one of greatest topographic significance consists of extremely large valleys, up to 7 km deep, with flanking rises reaching to within 1 km of the ocean surface, coincident with the large transform faults that offset the Mid-Atlantic Rise, throughout what has been called the Atlantic Equatorial Megashear Zone (Fig. 7.1). As such features are associated with a very prominent Mid-Atlantic Ridge the resulting topography is of such scale and ruggedness as to be hard to match anywhere else on the Earth's surface. As these features are developed along conservative plate boundaries there is at present considerable difficulty in trying to explain

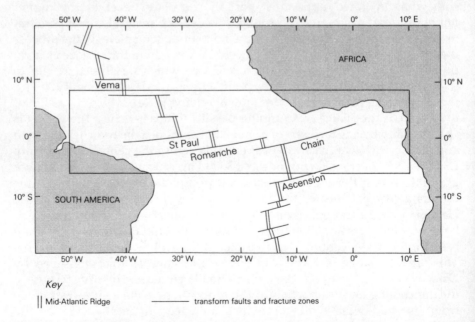

Key

|| Mid-Atlantic Ridge —————— transform faults and fracture zones

Figure 7.1 Transform offsets of the Mid-Atlantic Ridge (box encloses megashear zone).

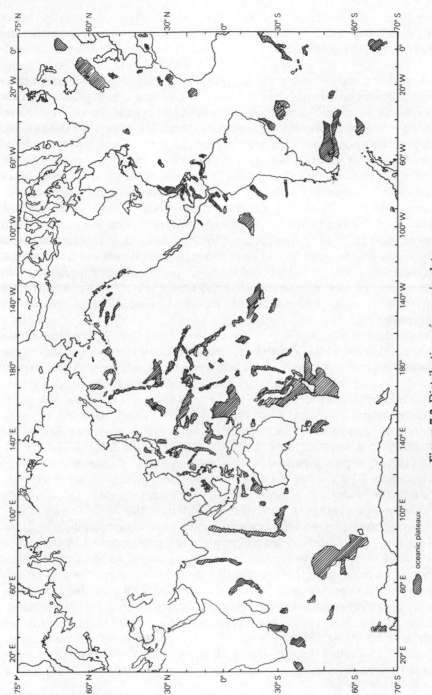

Figure 7.2 Distribution of oceanic plateaux.

oceanic plateaux

their formation in terms of plate tectonics. Possible solutions are being sought in the less-than-perfect operation of plate tectonics on or about the ridge offsets, but nothing that is really satisfactory has emerged so far.

Another group of features that are at present less than satisfactorily explained are the oceanic plateaux (Fig. 7.2). They rise to a considerable height above the general level of the abyssal plain and possess crustal thicknesses much in excess of what is normal for oceanic lithosphere. They occur for the most part in the western Pacific and Indian Oceans and, although compositional data are sparse, there are enough to show that many are purely oceanic, being fossil island arcs and linear island chains. However, there are a few that consist of continental materials; Precambrian granites are exposed in the Seychelles and have also been dredged from the Agulhas Plateau. There are also indications that continental material is involved in the Lord Howe Rise, east of Australia, and the Ontong–Java Plateau, east of New Guinea. A great deal more work is required before it will be possible to reach conclusions about the origin of these features, but the solution to this problem is becoming urgent in view of recent work in North America (see later in this chapter) where it has been demonstrated that these oceanic plateaux are an essential component of continental orogenesis.

In contrast it is possible to offer a fairly complete explanation of the continental shelf, slope and rise and why they are largely confined to the Atlantic and Indian Oceans, while being practically absent from the Pacific. However, as this explanation is so intimately associated with the continents and continental movements it will be best to discuss it in the next section.

At the beginning of the chapter, it was mentioned that a big problem involved in determining the origin of continental topography lies in the possibility of inheritance from previous cycles of plate tectonic activity. To avoid this, it is proposed to confine attention to the effects of the present cycle initiated by the break-up of Pangaea (see Ch. 6). These effects are located either along the external margin of the supercontinent or along the internal margins resulting from its fragmentation, the former being associated with converging and the latter with diverging plate margins (Fig. 2.6).

However, contemporary diverging plate margins, marked by the development of the mid-oceanic ridge, are now confined to the oceanic lithosphere. The margins of the Atlantic and Indian Oceans are the sites of past continental divergence and the continental materials involved are now deeply buried beneath sediments that characteristically form the continental rise, slope and shelf. These generally level-bedded sediments are underlain by more laterally variable sedimentary and volcanic rocks, deposited in faulted rifts in the underlying thin continental basement, which is the original internal continental margin developed during the disruption of Pangaea.

It is now possible to interpret such a sequence in terms of plate tectonics. Initial plate divergence beneath the supercontinental masses resulted in

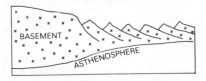

Figure 7.3 Thinning of continental lithosphere on a passive margin.

the thinning and attenuation of the overlying continental lithosphere by the development of normal curved (listric) faulting (Fig. 7.3), accompanied by basaltic volcanism, resulting from the partial melting of the closely underlying hot mantle material. The East African rift valleys seem to be a present-day example of this stage of continental disruption. Continued plate divergence eventually resulted in the splitting of the supercontinent, with the initiation of an intervening ocean basin and the inactivation of the faults associated with the previous stretching and attentuation of the continental lithosphere. The Red Sea provides a present-day example of such an early stage in ocean basin development. Further movement of continental margins away from the zone of divergence led to their slow and gradual submergence, because of the associated cooling and contraction of the underlying plate. In turn this led to the slow and long-continued deposition of material eroded from the continents (**miogeoclinal deposition**) giving rise eventually to today's continental rises, slopes and shelves, which are such a feature of these internal or passive margins. Thus a major oceanic feature is largely explicable in terms of processes affecting the continental lithosphere.

As the offshore zone was depressed and submerged, the coastal area was uplifted; this can be seen on either side of the Red Sea, where hills rise relatively steeply from the coast and then slope gradually down into the interiors of Arabia and Egypt. This is developed on a much grander scale, often with very considerable basaltic volcanism, around the shores of the Atlantic and Indian Oceans (Natal and south-eastern Brazil) and also along the east coast of Australia. The main features of these coasts were for the most part blocked out as a result of the Mesozoic fragmentation of Pangaea, which is to be contrasted with the situation on active margins, where most features are no older than early to middle Tertiary in age.

Despite the fact that continental lithosphere is involved, the effect of the tensional faulting resulting from plate divergence to produce passive margins is localised and is never more than a few hundred kilometres across. It is only when the continental lithosphere is under compression, as it is on external or active margins, that reactions to plate movement become spread over a very considerable area. Nowhere is this effect better seen than in the collision between India and Eurasia, where, it has now been suggested, marked topographic changes have been produced at great distances from the actual zone of collision. However, this solution has only

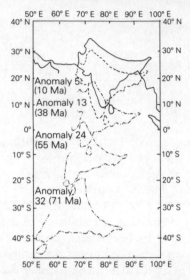

Figure 7.4 Movement of India according to oceanic anomaly pattern.

been reached as a result of the quantification of continental movement made possible by plate tectonics.

The key to this quantification is the magnetic reversal pattern of the ocean basins. By plotting those of the North Atlantic over time the positions of Eurasia and Africa with reference to North America were established. Similar work in the Indian Ocean located the positions of Africa and India (see Ch. 6 for dating of magnetic anomalies). It was then possible to locate India with respect to Eurasia at various times in the past (Fig. 7.4).

This shows that the rate at which India is moving, relative to Eurasia, has decreased considerably over the past 40 Ma. By plotting how far over time the northeastern and the northwestern extremities of India have differed from their present positions, it can be seen (Fig. 7.5) that the continent was moving at 100–180 mm a^{-1} and this decreased to about 50 mm a^{-1} around about 40 Ma ago. The sudden decrease in the rate of movement has been taken to mark the time at which India first collided with Eurasia, a contention supported by other lines of evidence. A sequence of sedimentary rocks has been described on the northern side of the Himalayas, typical of those deposited on passive continental margins, i.e. miogeoclinal, which began to be formed in the Cambrian, with deposition ceasing in the Eocene. This points to an ocean existing between the converging continents until at least 55 Ma ago. Evidence of actual continental collision is provided by the sudden appearance in India of Mongolian related mammals 45 Ma ago, whereas prior to that date no fossil mammals of any kind have been found.

Before this collision, of about 40 Ma ago, it is envisaged that the northern

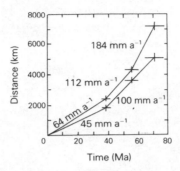

Figure 7.5 Change in rate of movement of India.

part of the Indian Plate, formed of oceanic lithosphere, was being subducted beneath Eurasia in a relatively smooth and regular manner. This supposition is supported by the presence of andesitic volcanics in southern Tibet, which is the expected result from the partial melting of a subducting oceanic plate. In addition, at the junction between the two continental masses, along the Indus and Tsangpo Valleys in southern Tibet, a fragment of this oceanic plate has been preserved. These rocks, ophiolites (referred to in Ch. 6), contain representatives of each of the three oceanic layers (see Ch. 4): bedded cherts formed as deep sediments, pillow lavas erupted on or about a mid-oceanic ridge, and ultrabasic rocks from the mantle. During the period when the oceanic plate was being subducted, the land areas of both India and Asia were tectonically stable platforms of low elevation. There is no evidence of large-scale vertical motion to produce the present-day relief before the Oligocene, that is, before the collision occurred between the continental masses.

India was undoubtedly slowed down by the collision, but even 50 mm a^{-1} is a very considerable speed in continental terms, sufficient to move the continent 1500–2000 km relative to Eurasia in the 40 Ma that have since elapsed. It was thought originally that this movement could be accounted for by localised lithospheric shortening connected with the formation of the Himalayas and underthrusting by India to form the Tibetan Plateau, but recent work has suggested that this solution is much too parochial.

There is no doubt that the Himalayas are the result of lithospheric shortening, for they consist of slices of the original northern margin of India stacked one on top of another. The faults defining these slices are of decreasing age in moving from north to south, reflecting the gradual penetration of the stress of the collision further and further into India. The overall effect of this stacking is to double the thickness of the lithosphere across a zone some 300 km wide, which indicates a shortening of the same amount.

The idea of massive underthrusting to explain the high altitude of the

125

Figure 7.6 Tectonic features of eastern Asia.

Tibetan Plateau is, however, very suspect, for its presence should be marked by a very long (about 1000 km), very shallow dipping (5° or less) fault zone and there is no evidence for such a structure ever having developed. Present ideas for the high altitude of Tibet centre on it being the result of thermal processes in the lower part of the lithospheric plate causing expansion and elevation of the surface.

Within Tibet the only effect that can be ascribed to the collision is a certain amount of near-surface folding and thrusting, largely confined to the south of the country, but even this could possibly date from a period prior to the collision.

Therefore, to account for the major part of the post-collision movement it is necessary to look beyond the immediate zone of continental suture, along the Indus–Tsangpo Valleys, at structures throughout eastern Asia. The most outstanding of these are the mountains of central Asia, the Pamirs, Tien Shan, Nan Shan and Altai, which reach heights of over 6000 m. In all of these, thrust structures are dominant, leading to considerable lithospheric thickening (Fig. 7.6).

That the greater part of this lithospheric thickening, connected with the elevation of these mountains, is fairly recent is indicated by the work in the Tien Shan of Russian geologists, who have shown that the topography in this area was very gentle from 200 Ma to 40 Ma ago and it is only since then, i.e. since the India–Eurasia collision, that the mountains have been elevated. Other work has shown, not only in the Tien Shan but also in the Pamir and Altai ranges, that there was a vertical elevation of 2–5 km in the late Tertiary. It would seem, therefore, that these central Asian mountains have been produced by the transmission of compressive forces through the continental lithosphere for up to 2000 km as a result of the continued northward movement of India. The amount of shortening involved in this mountain building would seem to be of the order of 200–300 km, so there still remains a considerable amount of the post-collision differential movement between India and Eurasia to be accounted for, possibly somewhere between 500 and 1000 km.

A solution was suggested by ERTS imagery of central China, which shows a series of long linear valleys and adjacent ridges that were recognised as being the traces of active transcurrent or strike-slip faults (Fig. 6.6). Although some of these faults were known from displacements associated with earthquakes and from geological studies, others were unknown until the satellite imagery became available. By combining these topographic data with fault-plane solutions of earthquakes and observed surface displacements following some of the largest earthquakes, it has been possible to determine the direction of motion of the main transcurrent faults (see box). The pattern that emerged is quite simple (Fig. 7.6); between the Pamirs and the Tien Shan the faults such as the Talasso-Fergana have a NW–SE strike and a right lateral motion. In China and Mongolia the faults such as the Altyn Tagh and Kunlun have a generally east–west strike with a left lateral motion. This is the pattern to be expected if pressure is applied to the region in a direction between north–south and NE–SW, such as would be caused by a collision between India and Eurasia. Moreover, the predominance of this transcurrent faulting over thrust faulting suggests that the northward motion of India is accommodated mainly by material moving out of the way of the impinging continent. However, there is little or no information regarding the degree of displacement along these faults, or the time at which such movements have occurred. It is possible by using rather more indirect means to obtain some idea of the extent of the displacement. Thus in the case of the Altyn Tagh Fault, restoration of 400 km of left lateral motion would place the Altyn Tagh alongside the Nan Shan Mountains and open a connection between the Tsaidam and Tarim Basins. While such evidence is in no way conclusive when it is combined with the prominence and continuity of the fault on the ERTS imagery (which is at least as good as that of the San Andreas Fault that has moved 300 km since the early

Miocene), it becomes reasonable to conclude that movement along these faults of probably hundreds of kilometres has occurred. The widespread nature of seismicity in Asia also indicates the current nature of this movement. Not only are there numerous small earthquakes north of the Himalayas, but of the seven great earthquakes to have affected Asia since 1897, four occurred north of the Himalayas, as did more than half of the 75 earthquakes with a magnitude greater than 7 that occurred in the same period.

Direction of motion of transcurrent faults

Strike-slip or transcurrent faults are classified by considering the fault movement as if you were standing on one side of the fault and looking across to the other. If the other side moves to the right, it is a right lateral fault; and to the left, it is a left lateral fault. The fault shown in Figure 6.6 is a right lateral fault, for no matter which way it is viewed the other side is moving to the right.

In general, the clearest, and therefore presumably the most important, faults are the left lateral faults that run east–west, compared to the NW–SE right lateral faults. This must mean that most of the lateral displacement of China out of the way of India is to the east. This is understandable; material displaced to the west would have to push against thousands of kilometres of the Eurasian land mass whereas movement to the east would not be so impeded for material would be able to thrust over the oceanic plates along the margins of the Pacific. Thus it is possible that the whole of south-east China has been moved to the east for a distance anywhere between 500 and 1000 km along these faults. From this, it would seem that transcurrent faulting can be of great importance in continental areas, while in oceanic basins transform faults are the dominant form.

As well as reacting to the Indian impact by lateral movement along strike-slip faults, Eurasia in areas further to the north-east is also being wedged apart. Thus Lake Baikal occupies a tensional rift in which the normal faults are trending NE–SW and there is also a complex zone of rifting within the great northern loop of the Hwang-Ho in eastern China, the so-called Shansi Graben System. The important point about these two rifted areas is that they are now interpreted as being the result of the India–Eurasia collision and, therefore, are not related to forces operating directly below the rift but to forces lying thousands of kilometres to the south-west. (It is possible that the Rhine Graben north of the Alps can similarly be ascribed to the northward drive of Africa rather than to any diverging mantle currents operating directly beneath.)

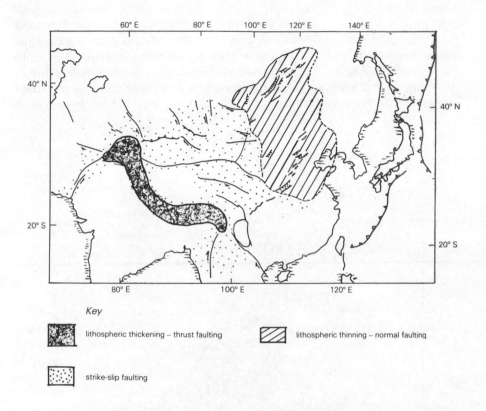

Key

lithospheric thickening – thrust faulting

lithospheric thinning – normal faulting

strike-slip faulting

Figure 7.7 Tectonic style in eastern Asia.

As a result of this work there is now emerging a rational explanation of the macrotopography of eastern Asia, for it is possible to differentiate three very different tectonic styles (Fig. 7.7): major lithospheric thickening, strike-slip faulting with lithospheric thinning, each of which gives rise to very different topographies. In addition, this work is having a profound effect on ideas concerning mountain building, for it has shown that orogenesis results from a collision between two continental masses, but that it does not occur when an oceanic plate is being subducted beneath a continental one; for no mountain building occurred in southern Asia in the Cretaceous when only the oceanic part of the Indian Plate was being subducted, before the actual continental collision. Yet, the origin of many of the circum-Pacific mountain chains, such as those of the west coast of North America and particularly the Andes in South America, has been ascribed to just such a process, i.e. the subduction of an oceanic plate beneath a continental one. This has presented Earth scientists with a major paradox;

recent work on the west coast of North America points to a possible solution.

In the late 1960s, at the time the idea of plate tectonics became generally acceptable, the west coasts of North and South America were regarded as being very similar, with the Pacific oceanic plates being subducted beneath the continents. South America provided the most straightforward example, for in the North American case the continental plate in moving west had practically overwhelmed the subduction zone. In both cases, however, it

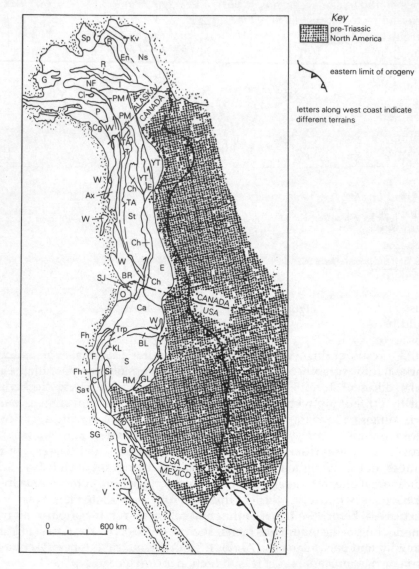

Figure 7.8 Areas added to western North America since the Mesozoic.

was accepted that volcanism and orogenesis were connected, with both being related to the subducting oceanic plate, and that additions to the continental mass from the oceanic plate, other than by volcanism, were of minor importance.

In the subsequent investigation of the plate tectonic history of North America, it became evident that from late Precambrian to late Triassic times, as Pangaea was being formed and a proto-Pacific developed, the western edge of the continent was a passive margin, across which was deposited a series of miogeoclinal sediments derived from continental sources. With the break-up of the supercontinent beginning in the late Triassic, this western margin became active, with the development of subduction zones, volcanism and eventually orogenesis. However, despite all this subsequent activity it is still possible, from the characteristics of the miogeoclinal sediments, to map the original western boundary. When this was done (Fig. 7.8) it was realised that a huge area of western North America had been added to the continent since the western margin assumed an active character in the late Triassic. This was very considerably at odds with the idea of gradual and minimal additions to the continents, for in this case over a period of something less than 200 Ma western North America had increased in area by some 25%.

Another problem was the source of this additional material. Strontium isotope ratios indicated a considerable change across this old western boundary, as defined by sediment type. The miogeoclinal sediments have a high $^{87}Sr/^{86}Sr$ ratio, confirming their continental derivation, while in materials to the west the ratio is lower, indicating strong oceanic affinities. More detailed investigation of these post-Triassic additions showed that most of the materials involved had originally formed parts of island arcs, oceanic crust and their associated sediments, with only a small proportion that could be described as being continental. These rocks now occur as a vast mosaic or collage of lithospheric blocks, which form separate geological entities, or terranes, each of which has a distinctive rock sequence differing markedly from that of its neighbours and separated from them by major faults. Palaeomagnetic determinations from individual terranes indicate that they have migrated thousands of kilometres from the south, over a period of tens of millions of years, with inferred velocities of about 5 cm a^{-1} before reaching North America. In other words, these terranes consist of the oceanic plateaux referred to earlier in this chapter (Fig. 7.2) and the absence of these features from most of the eastern half of the Pacific would seem to indicate this as their source area.

Despite their strong oceanic affinities, these exotic terranes were not subducted when they arrived at the junction between the Pacific and North American plates, but instead throughout the Mesozoic and early Tertiary they collided with continental North America, which was moving west as a result of the development of the Atlantic Basin.

The terranes were subject to intense thrust faulting and the detached sheets so produced were carried eastwards, initially, in mid-Jurassic times, over the continental margin and later, throughout the Cretaceous and into the early Tertiary, over the older thrust sheets, until now they form a belt of new continental lithosphere up to 600 km wide. In the latter part of this period, the thrust faulting was accompanied by extensive strike-slip faulting, as if the terranes were moving out the way of the advancing North American plate, in a similar manner to Eurasia's reaction to India's advance. The results of this can be seen in the way that particular terranes have been drawn out into thin strips parallel to the continental margin (Fig. 7.8). In more general terms this faulting has been responsible for shifting large parts of the Canadian cordillera hundreds of kilometres northwards with respect to North America as a whole.

The westward-moving continent was affected by a period of intense and widespread deformation and mountain building, as far east as Denver, between 40 and 80 Ma ago (Fig. 7.8), producing the giant uplifts of the Rockies and the Colorado Plateau, which together give the cordillera its extraordinary width. Most of the deformation of the Canadian Rockies and of the Sierra Madre Oriental of Mexico also occurred at the same time.

Previously, two ideas had been advanced to explain this orogeny. The first depends on the subduction of the Pacific plates off the coast of California and Oregon. To explain how this process could be the cause of mountain building more than 1200 km to the east in Colorado, it was postulated that the subduction angle of the oceanic plate was so shallow that even so far inland as Colorado it was still coupled to the overlying plate thrusting it upwards. The second possibility was that the North American and the oceanic plates were moving towards each other at such a speed that deformation took place at an unusually great distance from the subduction zone. Although there is some evidence to support both contentions, many feel that even if both mechanisms were working together they would not be adequate to account for such a wide and profound orogeny. However, the concept of the collision with accreted terranes appears to provide a more-than-adequate cause for this major orogeny.

This idea has progessed to such an extent that the classical explanation of the Andean orogeny, in the sense of the subduction of normal oceanic lithosphere beneath a continent, is now being vigorously questioned. It seems that, wherever enough structural, stratigraphic and palaeomagnetic data are available, the presence of accreted terranes can be established and that orogenesis is intimately linked with the incorporation of these terranes. As yet, no sufficiently detailed work has been done in the Andes to establish whether exotic terranes are present or not, but already enough has been accomplished to suggest that the orogenic history is not as simple as that to be expected from straightforward subduction.

This is supported by the fact that off the west coast of Central and South America there is strong evidence of contemporary terrane collision, for here there are four ridge-like oceanic plateaux of probable volcanic origin

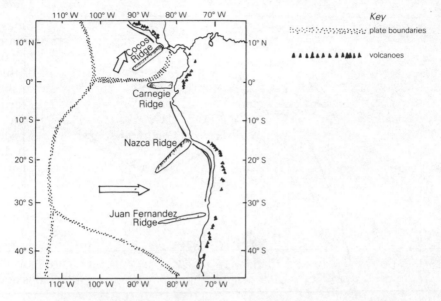

Figure. 7.9 Present-day collisions and their effect on volcanism.

impinging on the coast (Fig. 7.9). As they are all in isostatic equilibrium, having deep, light roots, it is reasonable to assume that they are not being subducted. Furthermore, each of them is having an effect on the seismicity, volcanism and morphology of adjacent areas. Thus, at the point where the Nazca Ridge abuts against South America, the trench is greatly diminished in depth and for 1500 km north of this point there is a gap in present-day Andean volcanism. This kind of evidence indicates considerable contemporary interaction between these oceanic ridges and continental orogeny.

However, a great deal more than the effects of contemporary collisions can be obtained from such investigations, for as oceanic plateaux are embedded in particular oceanic plates, they must move with the plate, so that if the speed and direction of plate movement are known, those of the embedded plateaux are also known. With these facts it is then possible to extrapolate plate motion backwards in time to investigate past collisions between oceanic plateaux and continents. By reversing the procedure it is equally possible to recognise where future collisions will occur.

Thus, in the Bering Sea during the late Mesozoic and early Tertiary (Fig. 7.10) subduction is thought to have taken place along the present-day continental margin. It would seem that this was brought to an end by the arrival of the Umnak Plateau at the subduction zone and its collision with the continent. The subduction zone was then relocated to the south of the plateau, resulting in the formation of the Aleutian Arc, a change that also isolated another northward-moving feature, Bowers Ridge, within the Bering Sea.

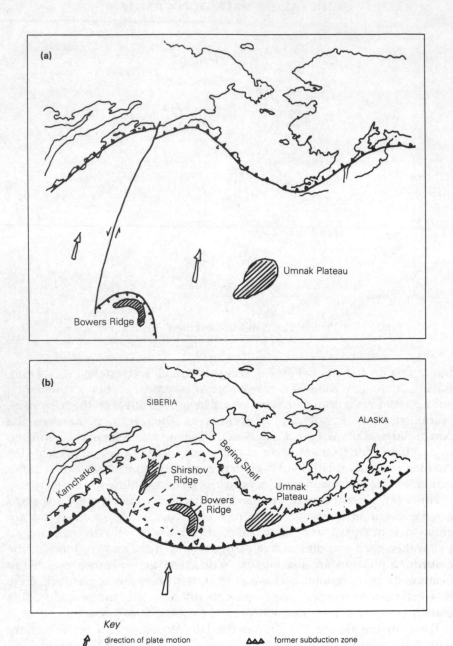

Figure 7.10 Postulated prior history of Umnak Plateau and Bowers Ridge leading to the development of the Aleutian Arc.

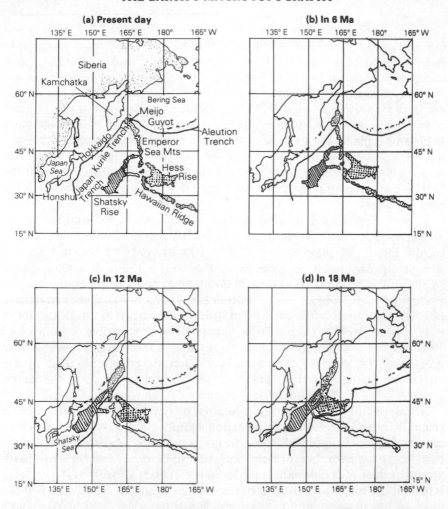

Figure 7.11 Possible development in the north-west Pacific.

A future collision is forecast for the north-west Pacific where the Emperor Seamounts are impinging on the Kurile Trench, with the Shatsky and Hess Rises somewhat further to the south (Fig. 7.11a). Based on present-day plate motions, it can be forecast that in 6 Ma (Fig. 7.11b) all these features will have moved to the north-west and that the Meijo guyot, after colliding with the subduction zone, will become part of the Kamchatka margin. In 12 Ma (Fig. 7.11c) the Shatsky Rise will collide with northern Honshu, Hokkaido and the Kuriles. At this stage the trench might move to the oceanic side of the rise to form to a new marginal sea, the 'Shatsky Sea'. In 18 Ma, all three features will be part of the Eurasian Plate, and new plate boundaries will be formed (Fig. 7.11d).

These three examples demonstrate the profound effects that oceanic plateaux can have not only on continental orogeny and continental margins but also on the process of subduction at compressive external margins. In more general terms the work on both diverging and converging plate margins is pointing to a much more intimate connection between things continental and things oceanic than was previously imagined and as a result of this breaking down of barriers a unitary theory of orogenesis is beginning to emerge. It is now possible as a result of the application of plate tectonics to give an explanation for most of the Earth's macrotopography. However, this is still a very single-faceted view, for there has been no overall reassessment of surface processes in terms of these changed views of the Earth's macrotopography. It is to be hoped that this will be remedied in the next few years.

As was emphasised previously, this is only one area from the many within Earth sciences that have been revivified and fundamentally changed by the impact of plate tectonics. The last 20 years has seen a science completely revamped as a result of the working out of plate tectonic implications and this process is still going on at an ever-increasing rate. In the history of the subject, the only other development that comes anywhere near it is the introduction by Lyell of uniformitarianism in the 1830s; hence the importance of the events that occurred in the Earth sciences from about 1950 onwards. By understanding them, it is hoped that there will be a greater appreciation of this particular scientific revolution and of the enormous strength and potential of this concept, which is the major achievement of 20th-century Earth sciences.

However, in view of the past history of Earth sciences, this state of scientific euphoria with regard to plate tectonics should be looked at with a certain degree of caution. Plate tectonics is now the new orthodoxy and is carrying all before it, but for how long will this continue and how will Earth scientists deal with heterodox opinion when it does appear, as it inevitably will. Is it too much to be hoped for that when this does occur they will not react in the unseemly and obscurantist manner that greeted Wegener's seminal speculations?

Suggestions for further reading

Asimov, I. 1967. *The universe from flat Earth to quaser*. London: Penguin.
Burchfield, J. D. 1975. *Lord Kelvin and the age of the Earth*. New York: Science History Publications.
Cohen, I. B. 1968. *The birth of a new physics*. London: Heinemann.
Hallam, A. 1973. *A revolution in the Earth sciences. From continental drift to plate tectonics*. Oxford: Clarendon Press.
Hurley, P. M. 1959. *How old is the Earth?* London: Heinemann.
Marvin, U. B. 1973. *Continental drift. The evolution of a concept*. Washington: Smithsonian Institution Press.
Sullivan, W. 1974. *Continents in motion*. New York: McGraw-Hill.
Wyllie, P. J. 1975. *The way the Earth works: an introduction to the new global geology and its revolutionary development*. New York: John Wiley.

The last chapter is concerned with much more recent work and hence reference must be made to papers in more specialist journals:

(a) On Asian orogenesis

Molnar P. and P. Tapponnier. 1975. Cenozoic tectonics of Asia: effects of a continental collision. *Science* **189**, 419–26.
Tapponier P. and P. Molnar. 1976. Slip-line field theory and large-scale continental tectonics. *Nature* **264**, 319–24.

(b) On North American orogenesis and oceanic plateaux

Coney, P. J., D. L. Jones and J. W. H. Monger. 1980. Cordilleran suspect terranes. *Nature* **288**, 329–33.
Ben-Avraham, Z., A. Nur, D. Jones and A. Cox. 1981. Continental accretion: from oceanic plateaus to allochthonous terranes. *Science* **213**, 47–54.

Index

Numbers in italics refer to text figure numbers. Numbers in bold type refer to boxes in text.